W. H. Stone

Elementary Lessons on Sound

W. H. Stone

Elementary Lessons on Sound

ISBN/EAN: 9783337277581

Printed in Europe, USA, Canada, Australia, Japan

Cover: Foto ©berggeist007 / pixelio.de

More available books at **www.hansebooks.com**

ELEMENTARY LESSONS

ON

SOUND.

ELEMENTARY LESSONS

ON

SOUND.

BY

DR. W. H. STONE,

LECTURER ON PHYSICS AT ST. THOMAS'S HOSPITAL.

WITH ILLUSTRATIONS.

London:
MACMILLAN AND CO.
1879.

The Right of Translation is Reserved.

LONDON:
R. CLAY, SONS, AND TAYLOR,
BREAD STREET HILL.

PREFACE.

The object of the present work, besides giving a concise outline of subjects required for examination, is to furnish information intermediate between Acoustics and Music proper, supplementary to both. Much of this material is only to be found in bulky, expensive, and foreign treatises; the rest is practically concealed in Memoirs and Transactions of Scientific Societies.

W. H. STONE.

CONTENTS.

INTRODUCTORY 1

CHAPTER I.

MODES OF PRODUCTION OF SOUND—VIBRATION OF SONOROUS BODIES 4

CHAPTER II.

MODES OF PROPAGATION OF SOUND—VELOCITY—WAVE-MOTION—REFLECTION—REFRACTION 46

CHAPTER III.

INTENSITY, CONSONANCE, INTERFERENCE 62

CHAPTER IV.

PITCH—ITS MEASUREMENT, LIMITS, VARIATION, STANDARDS, AND TONOMETRY 75

CHAPTER V.

NATURE OF MUSICAL TONE — QUALITY — HARMONICS — RESULTANT TONES 110

CHAPTER VI.

EFFECTS OF HEAT, ATMOSPHERIC PRESSURE, MOISTURE, DENSITY 127

CHAPTER VII.

SCALES, CHORDS, TEMPERAMENT, AND TUNING 134

CHAPTER VIII.

SPECIAL APPLICATIONS TO MUSIC—THE EAR AND VOICE . 161

LIST OF ILLUSTRATIONS.

FIG.		PAGE
1.	Vibrations of Stretched String.	7
2.	Harmonic Sounds. Nodes and Ventral Segments of a Vibrating String	8
3.	Harmonics. Nodes and Ventral Segments of a Vibrating String	9
4.	Sonometer	12
5.	Longitudinal Vibrations of Rods	16
6.	Marloye's Harp	17
7.	Vibrations of a Metal Rod	18
8.	Nail Fiddle	19
9.	Zanze	20
10.	Marimba	21
11.	A Tuning-fork mounted on a Sounding-box.	23
12.	Vibrations of a Plate.	25
13.	Nodal Lines of Vibrating Circular, or Polygonal Plates, according to Chladni and Savart	26
14.	Nodes and Segments of a Vibrating Bell	28
15.	Section of a Bell	32
16.	Proof of the Vibration of a Glass Bell	33
17.	Striking Reed	35
18.	Free Reed	35
19.	Egyptian Flute	37
20.	Prismatic Sonorous Pipes	38

LIST OF ILLUSTRATIONS.

FIG.		PAGE
21.	Cylindrical Sonorous Pipes	38
22.	Trevelyan's Instrument	41
23.	,, ,, Cause of Vibratory Movements.	42
24.	Philosophical Lamp or Chemical Harmonicon	43
25.	Kundt's Tube	56
26.	Experimental Study of the Laws of Reflection of Sound	57
27.	Sonorous Refraction. M. Sondhauss's Instrument	58
28.	Propagation of a Sonorous Wave through an Unlimited Medium	63
29.	M. Helmholtz's Resonance Globe	66
30.	Beats of Imperfect Unison. Ordinates of Ten Waves	71
31.	Ordinates of Eleven Waves transmitted in the same time	71
32.	Sums of the Corresponding Ordinates	73
33.	Savart's Toothed Wheel	78
34.	Seebeck's Siren	80
35.	Cagniard de Latour's Siren	81
36.	Interior View of the Siren	81
37.	Helmholtz's Double Siren	82
38.	Vibroscope	86
39.	Combination of two Parallel Vibratory Movements	88
40.	Vibrations of Compound Sounds	89
41.	Optical Study of Vibratory Movements	90
42.	Optical Curves representing the Rectangular Vibrations of Two Tuning-forks in Unison	91
43.	Combination of Two Rectangular Vibratory Movements	92
44.	Optical Curves. The Octave, Fourth, and Fifth	93
45.	Open Tube, with Manometric Flame	94
46.	Apparatus for the Comparison of the Vibratory Movements of Two Sonorous Tubes	95
47.	Manometric Flames. Fundamental Note, and the Octave above the Fundamental Note	96
48.	Manometric Flames simultaneously given by Two Tubes at the Octave	97

LIST OF ILLUSTRATIONS.

FIG.		PAGE
49.	Manometric Flames of Two Tubes of a Third	98
50.	Professor Blake's Method of Photographing Vibrations	99
51.	Curve representing a Sound-wave	111
52.	Perronet Thompson's Keyboard	147
53.	Poole's Keyboard	151
54.	Bosanquet's Generalised Keyboard	153
55.	Plan of Natural Fingerboard	157
56.	Organ Stops	165
57.	The Flute. Longitudinal and Transversal Section of the Mouthpiece	172
58.	Nay or Egyptian Flute	173
59.	Hautbois. Front and Side View of Reed	175
60.	Clarinet	175
61.	Section of Mouthpiece	176
62.	Bassoon	176
63.	French Horn	177
64.	Trumpet and Clarion	178
65.	Trombone	179
66.	The Human Ear	181
67.	The Human Voice	186

ON SOUND.

INTRODUCTORY.

Sound, from a *physical point of* view, may be defined as *Vibration appreciable to the ear.* It appears, at its lower limit, to be continuous with vibration as detectable by ordinary tactile sensation; hence its **exact** musical commencement **is** rather indefinite. It is usually **given** at about thirty-two **single,** or sixteen **double** vibrations per **second.** **Apparatus for the** demonstration of this **fact** will be **noted further on.** Its higher **limit** is even more variable, **owing** to physiological differences between different ears; **but** 76,000 **single,** or 38,000 double vibrations probably represent the highest note ever **heard.**

The line of demarcation between mere *noise* and *musical sound* seems similarly vague. Dr. Haughton has ingeniously shown that the rattling of vehicles over equal-sized stones becomes **musical at a definite velocity;** from the confused rattle of **a railway train in a tunnel the practised ear can** disentangle, **and, as it were** mentally sift **out,** grand organ-harmonies; **and** a falling plank in the Crystal Palace gives musical **notes** by periodic repercussion at equal intervals. On the **other** side, castanets, **tom-toms,** side-drums, triangles, cymbals, **all instruments of** music proper, only give noise, similar **to the guns** added in Russia to Italian music, or the hundred **anvils** "played **on"** at **the** Boston Celebration. **Even Beethoven,** in his grandest **symphony,** sounds every **note of the scale at once** with musical effect. Helmholtz **lays down the axiom that** the sensation of musical sound is caused by rapid **and** periodic movements of the sonorous body, the sensation **of** noise by non-periodic movements.

The present text-book is concerned mainly with the instrumental appliances which minister to sound. The physical aspect of Acoustics has been lucidly mapped out by Clerk-Maxwell in the following manner:—

VIBRATIONS AND WAVES.

Physical Aspect of Acoustics.

1. **Sources**—Vibrations of various bodies:—

Air	{ Organ pipes, resonators, and other wind instruments.
Reed instruments . .	The Siren.
Strings	Harp, &c.
Membranes	Drum, &c.
Plates	Gong, &c.
Rods	Tuning-fork, &c.

2. **Distributors**:—

Air	Speaking-tubes, Stethoscopes.
Wood	Sounding Rods.
Metal	Wires.

3. **Pugging of** floors, Dampers of **Pianofortes.**

4. Reservoirs, Resonators, **Organ Pipes,** Sounding-boards.

5. **Regulators** . . . . . Organ Swell.

6. **Detectors** . . . { The Ear, Sensitive Flames, Membranes, Phonautographs, &c.

7. Tuning-forks, Pitch-Pipes, Musical **Scales.**

To bridge over the gap between this science and the art of music, slightly more extended treatment is needed:—

1. MODES OF PRODUCTION. VIBRATION OF SONOROUS BODIES.

2. MODES OF PROPAGATION. **VELOCITY.** WAVE-MOTION. **REFLECTION.** REFRACTION.

3. INTENSITY. CONSONANCE. INTERFERENCE.

4. PITCH. MODES OF DETERMINATION AND MEASUREMENT. STANDARDS OF PITCH.

5. NATURE OF MUSICAL TONE. QUALITY. RESULTANT TONES.

6. EFFECTS OF HEAT. ATMOSPHERIC PRESSURE. MOISTURE. DENSITY.

7. SCALES. TEMPERAMENT. TUNING.

8. THE EAR AND VOICE. SPECIAL APPLICATIONS TO MUSIC.

The whole subject naturally and more concisely divides itself into—1. *Mechanical*; 2. *Theoretical* considerations; 3. *Practical* applications. It is, however, to be noted that these different aspects of the facts cannot be separated one from the other; the respective influences of the art of music and of scientific research having been reciprocal, gradual, and intimately combined. At the very outset of history we meet with the *Monochord*, named after Pythagoras, a machine not intended for artistic performance, but which at once yielded immense practical results to music. We then enter into a long period during which instrumental appliances grew, without design and without theory. The discoveries of new or improved instruments were purely technical, often fortuitous; although every instrument added was a piece of mechanism open to scientific analysis. During the present century a return has been made to apparatus essentially scientific, for the explanation of what had been musically invented; we find soon the reciprocal influence of instruments on apparatus—no better instance of which can be given than the discovery of Tartini's "*Terzo Suono*," or third sound; originally taught by the great violinist to his pupils as a means of accurate tuning, but now shown by Helmholtz to involve a new and important acoustic principle.

In many cases instruments of music actually stand in the place of apparatus. Strictly considered, a musical note is of itself a mathematical fact, quite independent of its power of exciting emotion and pleasure by its artistic production. On the other hand, tuning and intonation, originally left entirely to the accurate and cultivated ear of a skilled performer, have become a branch of science, with definite laws and practical rules; insomuch that the unconscious departures from a fixed tuning, which older musicians made by a kind of instinct, are now explained; and even the disposition of various instruments, with different qualities of tone, in an orchestra is shown to be correct, or the contrary, according as the harmonics of each peculiar quality are consonant or dissonant.

CHAPTER I.

MODES OF PRODUCTION OF SOUND. VIBRATION OF SONOROUS BODIES.

Sound may be generated in **various ways**; some of these have been utilized **for the** production of musical or regular vibrations, others remain in the **category** of mere noise.

The **shock** of two bodies against **one** another is perhaps the commonest of all sources, varying, however, materially according to the nature of the sounding masses and the **mode** of conveyance to the **ear**. The simple unmusical **tap of a** drumstick on **a** membrane is of musical **value in the side-drum, to** preserve time and indicate rhythm, while **in the** Morse telegraph the click caused by the collision of magnet and keeper is, by a suitable code, made **to furnish** an intelligible alphabet.

Irregular Vibration.

Friction in many **forms furnishes a source of** sound; the irregular vibrations, **which** in its rougher kinds it emits, passing imperceptibly **into** more regular and musical tones. This gradation may often be noticed in the sharpening of a saw, and still **more notably in the** action of railway brakes upon a train **in motion.** In its more refined adaptation to instruments **friction** originates **the tender** tones of the violin and its congeners; while in **the musical** glasses, **a** wetted finger moving with friction along the edge of **a consonant bell,** produces a quality of **sound almost painful** and **cloying** from its excessive sweetness. The friction of air against solid bodies originates **the melancholy moaning of the wind, the sharp** hiss of a **cane, and the loud genial crack of** a carter's whip.

Explosion as of gunpowder and **of explosive** gases, gives **rise** to loud noises; in the pyrophone **of** Wheatstone and others it has been made to serve, though as yet rather imperfectly, the function of a musical instrument. Of a kindred nature are the **various kinds of singing** and sensitive flames.

Electricity, whether in its disruptive discharge at high tension, causing the rolling of thunder, or in the less awful manifestations of frictional and inductive machines, is a source of sound. In the form of a dynamic current, it has also the power of producing such an alteration in the molecules of soft iron as to be accompanied by audible musical vibration. Latterly this has been adopted into the service of music in the various forms of telephone.

Regular Vibration.

These being some of the commoner and more physical sources of vibration appreciable by the ear, there remains a larger and more distinct class, which, from its special adaptation to the production of regular waves, furnishes the bulk of the instruments and contrivances used for eliciting pleasant tones, and for building up the æsthetical art of music. They will have next to be considered at some length, and may be enumerated as follows:—

Summary of Vibrations.

I. Of Strings { 1. Transverse, 2. Longitudinal, 3. Torsional } excited by { A. Plucking, B. Striking, C. Bowing, D. Impact of air } vary with { a. Nature of stroke. b. Place struck. c. Rigidity of string. }

II. Of Rods { 1. Transverse, 2. Longitudinal, 3. Torsional } with { A. Both ends fixed (approach to those of strings). B. One end fixed (nail fiddle, musical box). C. Centre fixed (tuning-fork). D. Both ends free, nodal points supported (harmonicon). }

III. Of Plates { 1. Radial, 2. Circular } As in Chladni's experiments (gong, cymbal).

IV. Of Bells { 1. Spherical { A. Excited by blows (clock chimes). B. By tangential or radial friction (musical glasses). } 2. Of complex figure (give compound notes with irrelevant harmonics). }

V. Of Membranes { 1. Independent (tambourines, zambomba). 2. With associated air-chamber (kettle drum, resonators). }

VI. Of Reeds { 1. Free (as in harmoniums). 2. Beating { A. Single (clarinet, organ reed). B. Double (bassoon, oboe). } 3. Membranous { A. The lips in brass instruments. B. The larynx (human voice). } }

VII. Of columns of air
- 1. Organ pipes
  - A. Open pipes.
  - B. Stopped pipes.
  - C. Half-stopped pipes.
  - D. Pipes with reeds.
  - E. Mixtures and mutation stops.
- 2. Consonance boxes and vessels.

VIII. Vibration caused by heat
- 1. Trevelyan's **rocker**.
- 2. Sondhaus's experiment.
- 3. Of flames
  - 1. Chemical harmonicon, pyrophone.
  - 2. Sensitive and singing flames.

IX. Caused by electricity
- 1. Current in iron bar, Reiss's Telephonic receiver.
- 2. Telephone, Microphone.

Strings.—Among the commonest and earliest modes of eliciting musical sounds may be named *strings*. They have contributed the largest share to instruments of all times and all countries, and were early employed for more accurate determinations. Their theory is simple comparatively to that of other sound-producers.

The string itself is supposed to be a perfectly uniform and flexible thread of solid matter, stretched between two fixed points. Although this ideal is not actually attained in practice, the deviations from it are not so great as to prevent necessary corrections being made. Its vibrations may be divided into *transverse*, *longitudinal*, and *torsional*, the former being the form more usually studied; in which, if the stretching due to lateral displacement be small in comparison with that to which the string is already subjected, it may be neglected.

Transverse Vibrations of Strings.

The laws are as follows:—

1. For a given string and a given tension, the time of a vibration varies directly, the vibration number inversely, as the length.

2. When the length of the string is given, the vibration number varies directly, the time of vibration inversely, as the square root of the tension.

3. Strings of the same length and tension vibrate in times which are proportional to the square root of the linear density, the vibration number being in inverse ratio to this.

The motions of a string thus fixed at its end and excited at some intermediate point, radiate from that point to the fixed

I.]   MODES OF PRODUCTION OF SOUND.   7

extremities, whence they are reflected in the opposite direction,

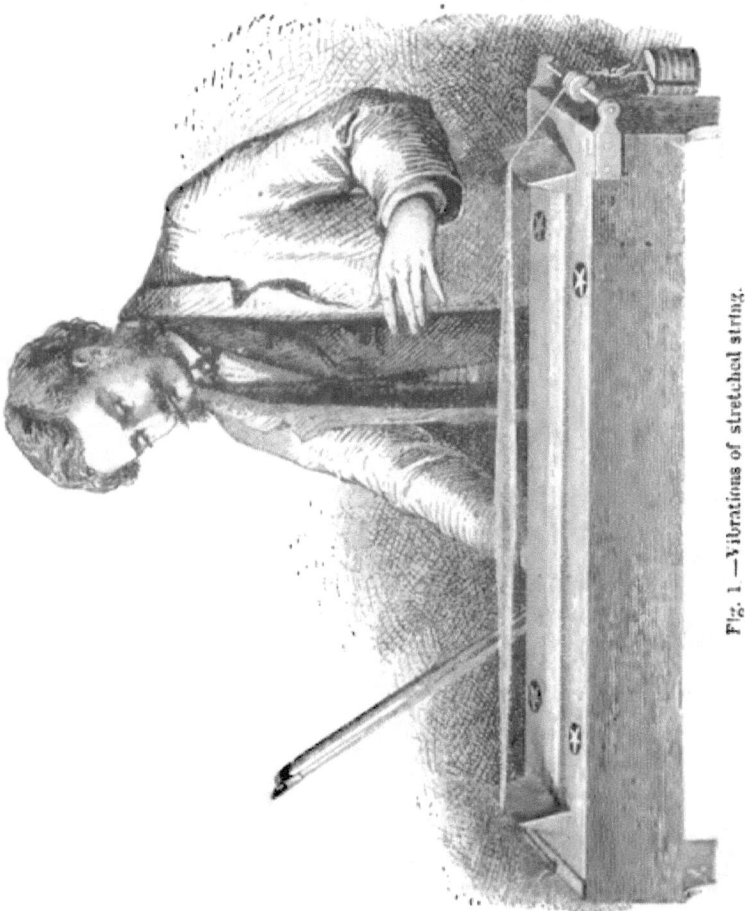

Fig. 1.—Vibrations of stretched string.

travelling over twice the length of the string.¹ The simplest

---

¹ The velocity with which transverse vibrations run along a flexible string is given by the formula $v = \sqrt{\frac{t}{m}}$, where t denotes tension, m the mass of unit length. The period of a complete vibration is therefore the time required for a pulsation to travel over twice its length: i.e. $n = \frac{1}{2 l}\sqrt{\frac{t}{m}}$; where l is the length of the string, and n the number of vibrations in a second, or the *frequency*. Period and frequency are therefore reciprocals.

form of vibration is that in which the string vibrates as a whole, and produces its lowest or fundamental note; but it may also be broken up into two or more *ventral segments*,

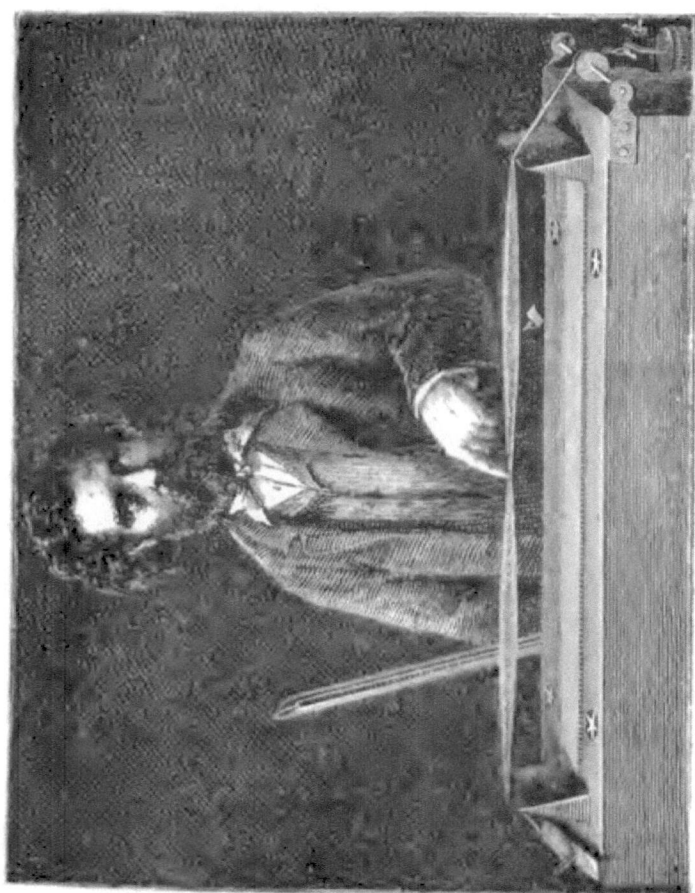

Fig. 2.—Harmonic sounds. Nodes and ventral segments of a vibrating string.

separated by *nodes* or points of rest, the rapidity of vibration being proportional to the number of these segments, and producing partial-tones, which will be described further on.

MODES OF PRODUCTION OF SOUND.

Nodes.

The breaking-up of the string into a number of *nodes* with intervening *loops* or ventral segments may also be determined with certainty by damping the exact point at which one of

Fig. 3.—Harmonics. Nodes and ventral segments of a vibrating string.

the nodes is situated, and exciting the string in the position of a loop or ventral segment. The corresponding nodal points throughout its length will thus be brought to rest, and the string will sound the upper partial tone belonging to the division. Thus if the damper be applied in the middle, a single nodal point will be formed, the string will vibrate in two halves, and give the octave of its fundamental note. If it be damped at a point one-third from the end, and excited midway between the end and the damping, a second node will be established in the free part; the string will vibrate in three segments, giving the twelfth of its fundamental note.

The higher notes thus obtained are termed harmonics, and will be considered later on. They can easily be shown by throwing a strong light on the string, or by means of small paper riders set astride on it, which are immediately thrown off at the loops, but not at the nodes.

Strings may be excited in various ways, and with corresponding variety in the sound produced. The oldest and simplest mode is obviously that of plucking them, drawing the tense cord out of its position of equilibrium and suddenly letting it go. This is the plan adopted in many ancient and modern instruments, such as the harp and guitar. Or the finger may be armed with a quill or plectrum, as in the case of the zither and in some oriental instruments. In the harpsichord this quill was inserted into a small moveable piece of mechanism termed the jack, which was itself actuated from the keyboard of the instrument.

A third and most important improvement is effected in the pianoforte, where a hammer of comparatively soft material strikes a blow on the string instead of twanging it like the harpsichord jack. But an entirely different course, early adopted, consisted in bringing a bow into frictional contact with the string, and by a succession of impulses conveyed to it, producing a continuous instead of an evanescent tone. In all the innumerable and ancient varieties of the viol and violin family, the bow is made of horsehair rubbed with rosin, kept at a moderate tension by the stick to which it is attached. Many attempts, only partially successful for the most part, have been made to adapt this method of excitation to instruments of the piano species, among which, elastic rollers, and rotating bows of parchment or horsehair passing over rollers, may be named.[1] A current of air directed against a string has long been known to be competent to excite it, and the Æolian harp has been constructed on this principle. Latterly attempts have been made to render this combination less

---

[1] The ancient "vielle," now degraded into the hurdy-gurdy, is the oldest, and perhaps the best representative of this class.

vague and fortuitous than it is in that primitive and intractable instrument, but hitherto without producing any very practical result.

It was shown by **Delezenne in 1842** that it is impossible to make a string sound **if it be excited in** the centre by a bow. Duhamel was of opinion **that** in **a string which is** giving its foundation **tone** the first partial **is vibrating also,** and **that since** the bow prevents this form **of motion, sound cannot be produced.** To verify **this** hypothesis, **he endeavoured to sound a string by means of two** bows **moving in the same direction, to the right** and left respectively **of** the **middle** point **of the** string. Still no sound was produced. But on the **other** hand, if **the** position of the bows were retained unchanged, and **an** opposite direction of motion with **equal velocity were given** them, the foundation tone **came** out **instantly,** accompanied by the first upper partial. If the **string be** attacked successively close to each of the consecutive **harmonic** points, **so to** produce the fundamental tone, the corresponding upper partial **is** reinforced. At one-third of the length the twelfth **has about** equal **intensity with the** fundamental, at **one-fourth the double** octave, **at one-fifth the** major seventeenth. The **harmonic** always slightly **precedes** the fundamental **tone.** Speaking generally, **a string vibrating** transversally can **only** sound **on** the condition **that it gives** two transversal **tones,** the **sharper** of which **depends on the point of attack, or the mode of excitement.**

Sonometer.

The **most** convenient apparatus for the performance **of** experiments on strings is the monochord or sonometer. This **is a** device of great antiquity. It consists, in its most modern **form, of a** long resonant box bearing on its upper surface **wrest-pins,** and two bridges set at a fixed distance, usually one **or two** metres apart. The space between these is occupied by **a** graduated scale, and a travelling bridge slides along the whole distance. By means of the wrest pins, one or more **wires** are strained over the fixed bridges **to** the **required** tension, and **any** given length **of string can be** cut off **by the** sliding bridge. Besides the wrest-**pins,** there is **usually at one** end **a** pulley grooved **to** receive **the** wire, **to which weights can be attached so as to verify the** second law given above. **A** modification of this latter contrivance, suited to bear a very considerable load, over two hundredweight in some instances,

12 ON SOUND. [CHAP.

yielded practical results in the hands of Colonel Perronet Thompson, and will be described in a subsequent chapter.

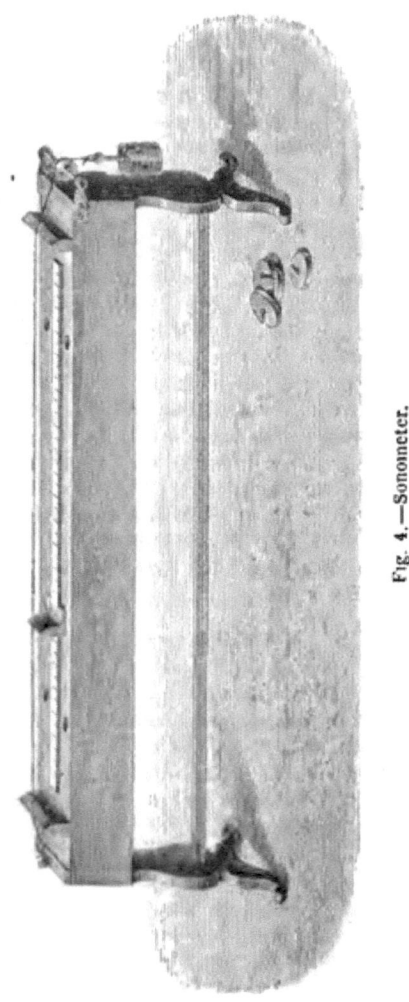

Fig. 4.—Sonometer.

Mons. Melde has introduced a striking method of illustrating the vibrations of a string, by fixing one of its extremities to the prong of a tuning-fork, and the other to a wrest pin, by means of which its tension can be varied. The fork being made to vibrate, and the string properly stretched, the latter oscillates in unison to the former in a manner easy to see from a distance. When the tension is gradually lessened, the single segment at first formed divides into two, then into three portions, each separated by motionless nodes.[1]

Longitudinal and Torsional Vibrations.

Longitudinal Vibrations.—Every string which vibrates transversely between two points must also vibrate longitudinally. This is evident from the fact that it cannot deviate from a straight line without lengthening, nor return to it without shortening. The resultant sound is usually difficult to hear in ordinary sonometers, because the two bridges at the extremities of the string being unable to stop the longitudinal vibrations, they are transmitted to the lengths of wire beyond, and thus stifled by interference. In the violoncello the effect of longitudinal vibrations is sometimes very unpleasantly sensible, forming one of the varieties of false note or "wolf."

Longitudinal vibrations have been little utilized in the production of musical sounds. Their frequency of vibration varies inversely as the length of string.

Even in the monochord a powerful acute sound can be obtained by rubbing the string lengthwise with a piece of rosined leather. By damping the string in the centre, a note is heard an octave higher, and by stopping it at one third, one a twelfth above the fundamental. These longitudinal vibrations are not affected by tension, as are the transverse, but are materially influenced by the substance of which the wire is made, and the velocity with which it transmits sound;[2] indeed, the wire in this form approximates in its behaviour to a rod or bar. In the case of the fundamental note each of the two halves of the string is alternately extended and compressed. At the middle point there is no compression but great amplitude of movement.

[1] A beautiful apparatus of this kind was exhibited at the Loan Exhibition, in which both ends of the string were connected with elastic bars kept in vibration by means of electro-magnets. The period and plane of the wave-motion could be regulated.

[2] Indeed the note emitted affords a good measure of the varying velocity of sound-waves in different materials. See the paragraph on velocity in solid bodies.

Sonometer for Longitudinal Vibrations.—A string to produce pure longitudinal vibrations must be stretched between heavy and firmly-fixed terminals. At each end of the long wooden resonance-box are clamps of solid metal faced with lead. A scale, usually of $1\frac{1}{2}$ metre or more in length, is placed below the string, which can be stretched either by wrest-pins or by a weight. A leaden clamp travels along the string. It is the **inertia** of **this weight** and that of the **terminals which determines the nodal points, and not** their rigidity as **in the case of transverse oscillations.** The vibrations themselves consisting of alternate rarefaction and condensation **of** the elastic material closely resemble those which take place in an organ-pipe, or in the **atmosphere** at large **when** conveying sound.

The string may be **excited by means of** the thumb and finger dusted with **powdered rosin, and** moved lengthwise, **or better** with **the point of a violin-bow** acting in the same direction.

The sounds thus **elicited** are very pure, and always much sharper relatively to **the** length of the string than those given by transverse vibrations. They have the same mutual relations **as** these **latter and** their vibration-numbers **are inversely** proportional **to** the wave-lengths.

Torsional Vibrations of Strings.—Independently **of** the two modes of vibration given above, **every** string **performs a** third oscillation inseparable from the **others.** This can easily **be demonstrated by** hooking lightly **to** the middle of the string **a small double** ring of fine **wire, in the shape** of the figure 8, carrying a little paper **flyer. When the** string **is** transversely excited, whether by plucking **or with a** bow, **the ring and flyer will begin to turn round** with **great** rapidity, alternating frequently **in the** direction **of rotation.** This occurs whether the **fundamental tone** or an upper partial be elicited. It is evident **that the string** has a torsional motion, **which it** communicates **with alternate** direction **to** the enveloping **ring.** The **torsional** vibrations **of strings** are of little **practical** importance. **But** they **unexpectedly** intervened in **the writer's** experiments **with low notes on the** double-bass **as** detailed elsewhere. Whenever **it was attempted to produce** grave **tones by** enlarging the sectional area **of the** string, a limit **was** found beyond which the diameter of **the** string could **not be** increased, from **the** predominance **of** these torsional tones. The bow, acting at the circumference of the string, had power enough to rotate it instead of communicating purely transverse oscillations. It was consequently necessary to adopt strings of smaller diameter and greater specific gravity.

It will thus be seen that a string vibrating in the transverse direction has impressed upon it at least four distinct motions, two in the original direction, one longitudinal, and one torsional. By accidental circumstances the number may be indefinitely increased.

Summary of String-Vibration.

I. *Transverse.*—Alone used in music.

Velocity $= \sqrt{\frac{t}{m}}$.

Period $= n = \frac{1}{2l}\sqrt{\frac{t}{m}}$.

The harmonics a complete series, **as in open pipes.**

II. *Longitudinal*—

Velocity $= \sqrt{\dfrac{\text{modulus of elasticity}}{\text{density of string}}} = \sqrt{\dfrac{M}{D}}$.

Unaffected by tension. Inversely as length.
Their pitch higher than in No. I. Vary in pitch with the material of the string.
Harmonics a complete series.
Afford a measure of sound-velocity,

III. *Torsional*—

Complicated in theory. Not used musically.

Vibrations of Bars or Rods are the next in simplicity to those of strings. They are of three kinds, *longitudinal, torsional,* and *lateral.* Of these the last are the most important. Although the three classes of vibrations are quite distinct in theory, yet in actual experiments it is often found impossible to excite longitudinal or torsional vibrations without the accompaniment of some measure of lateral motion. In bars of ordinary dimensions the gravest lateral motion is far graver than the gravest longitudinal or torsional motion.

Rods or bars with one end fixed can also be made to vibrate longitudinally, the pitch being inversely proportional to the length of the rod. The time of a complete vibration is that required for the sonorous pulse to run twice to and fro over the rod. The first upper partial of such a rod produces a node at one-third from its free end; the second has two nodes, the higher at one-fifth of the length from the free end, the lower bisecting the remainder of the rod. The order of the tones is that of the odd numbers 1, 3, 5, &c., thus resembling those of a stopped diapason pipe which will be described further on.

The only instrument founded on this property of rods is but little known and rarely used, being more an acoustical curiosity than anything else. It consists of a number of deal rods, about twenty or more, standing up vertically, like the strings of a harp, from a sound-board obliquely

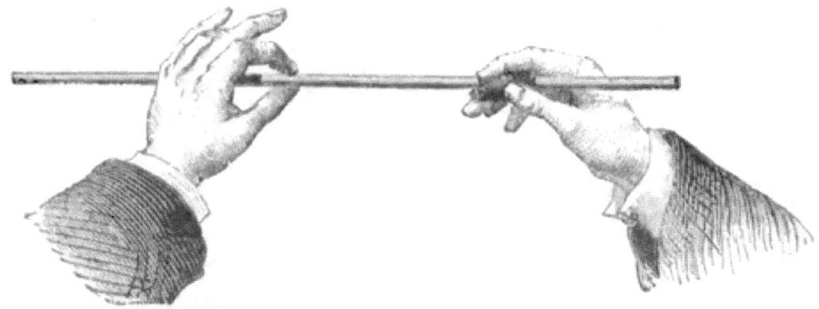

Fig. 5.—Longitudinal vibrations of rods.

placed below, into which their lower extremities are firmly fixed. They are excited by vertical friction with the rosined fingers. A similar instrument furnished with glass tubes instead of wooden rods is occasionally to be heard in the streets of London.

Torsional Vibrations are even of less acoustical importance than those named above. They were first, like them, investigated by Chladni, but they have hitherto contributed no instrument to music, and are chiefly of a mathematical and theoretical interest.

Lateral Vibrations of Elastic Rods, on the other hand, are of large service both theoretically and practically. They differ materially according to whether the rod is firmly fixed at one or both ends, or free at the two ends, and supported at some other point. A rod fixed at both ends behaves exactly like a string. It may vibrate in one, or in two, three, or more segments. But the rapidity of the vibrations, and the consequent pitch of the note produced, differ entirely from those of a string. Whereas the vibrations of the string rise in a simple arithmetical series, those of the rod rise as the squares of the odd numbers. When a string divides into two segments, each of these vibrates with twice the rapidity of the whole string; but a rod does so in the ratio of 9 to 25, a rod vibrating with two nodes in the ratio of 9 to 49, one with three nodes in that of 9 to 81, and so on. A rod supported at one end gives a number of vibrations in inverse ratio to the

Fig. 6.—Marloye's harp.

square of its length. It can be made to vibrate in a single piece, or with one, two, or even more nodes. The period of the gravest tone is the time occupied by a pulse travelling *four* times the length of the bar.

These properties of rods were utilized by Chladni in the construction of a tonometer.

Rods fixed at one end furnish several instruments of practical application, of which the *Nail-fiddle*, or *Violon de fer*,

18 ON SOUND. [CHAP.

seems the simplest. It is said to have originated by accident, from the note given by a common nail inserted in the wainscot when a weight was hung to it by a string. In this a bow is

Fig. 7.—Vibrations of a metal rod.

drawn over a graduated series of nails, or rods, fastened by one end to a block of wood, thus setting them in vibration. In the "Jew's harp" or *Guimbarde*, which in various forms appears in many parts of the world as a popular instrument, the fundamental tone of the spring is modified by alterations in the cavity of the mouth to which it is applied. A curious instrument of the spring class is the *Zanze*, from Western Africa, shown in the late Loan Exhibition at South Kensington.

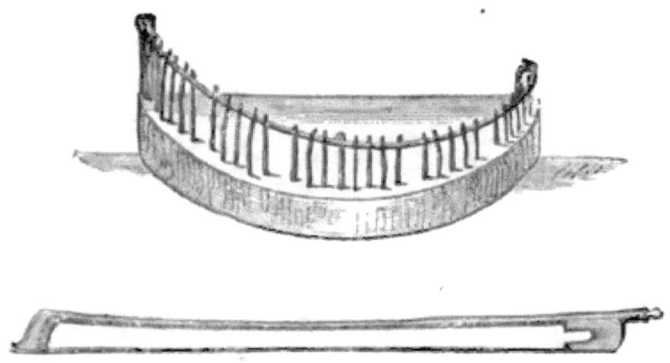

Fig. 8.—Nail-fiddle.

It consists of a carved wooden box, having at the top a number of iron tongues, which the performer sets in vibration by means of his thumbs. This approaches nearly to the *musical-box*, in which a "comb" of steel vibrators, weighted in the lower octaves with lead, are plucked by pins inserted in a revolving barrel. The "*gongs*" of American and other clocks are rods coiled into a flat spiral, attached to a heavy mass of metal at the fixed end, and firmly screwed to the wood of the case. An instrument termed the "*Bell Piano*," with single springs struck by hammers actuated from a keyboard, is also made. In *tuning-forks*, the necessity for firmly fixing the base of the rod or spring is obviated by attaching it to a second rod or spring, which vibrates in opposition, and keeps the whole mass in equilibrium.

Rods or Bars free at both ends may vibrate in several ways. In the simplest case the rod has two nodes, and three vibrating segments; the central segment is the longest, the distance of each node from the end being about one-fourth of the distance between them. In the second form the rod has three nodes, and four vibrating parts, the central segment dividing into two. The fundamental note of rods thus arranged is higher than that of a rod fixed at one end in the ratio of 4 to 25. The upper partials are not much required, the first mode of vibration being that commonly utilized.

These vibrations of rods rise to their highest musical character in instruments commonly called *Harmonicons*, which

Fig. 9.—Zanze.

may be made of wood, glass, steel, or even of compact crystalline stone. Many oriental specimens are made of the first material, usually of the siliceous outer layer of the bamboo, and are remarkable for the presence of resonators reinforcing the note, which will be adverted to farther on. The tone is astonishingly large and pure. Such an instrument of pine wood, bearing the name of *Xylophone*, has recently been produced at many London concerts. Another, made of compact slate in bars, the *Rock Harmonicon*, was exhibited a few years back. **Mozart writes for a similar** instrument, probably made of **steel,** in his **opera of the** *Flauto Magico,* where it is intended **to imitate the sounds of** the classical *Sistrum.*

Fig. 10.—Marimba.

By far the most important case of rods or bars free at both ends is that of the ordinary tuning-fork, named above.

The Tuning-fork, as an acoustical instrument of paramount interest, requires to be described in detail. It may be looked upon as an elastic bar, free at both ends, and supported in the middle where the stem is inserted, or as two mutually antagonistic bars, vibrating in opposite phases, so that the general centre of inertia is undisturbed. It is usually made of steel, but is equally efficient if formed of hard brass, or of the compound of tin and copper, called bell- or gun-metal.[1] In this case, the bar, instead of remaining straight, is bent until the two vibrating branches stand parallel to one another. Its pitch becomes somewhat flatter after this change, and the nodal points approximate to one another. The stem or handle is usually inserted into the convexity of the bend, and, in the best constructed forks, is spread into a solid block of metal continuous with the fork; indeed the writer has found the sound emitted by a fork in which the curved part is extended into a triangular prolongation fuller and more pure than in the ordinary construction. The support in this case stands at right angles to the plane of the fork's vibration, being inserted into a hole drilled at the centre of the triangular area.

It will be seen that part of the motion given to the ordinary tuning-fork is, by its shape, transmitted in the direction of the stem or handle, which has an up-and-down oscillation at right angles to that of the prongs themselves. This component can be farther transmitted to a resonant body, and the tone materially augmented by consonance.

The sound of a tuning-fork, when struck alone, contains, besides the fundamental note, numerous upper partial tones; but the interval between them and the lower sound is infinitely greater than in the case of strings. In those examined by Helmholtz, the number of vibrations of the first harmonic varied from 5·8 to 6·6 times that of the fundamental; the rates of the whole series being as the squares of the odd numbers 3, 5, 7, 9, &c. The result of this is that the upper partials are singularly evanescent, and soon leave the fundamental practically pure and uncomplicated. This important acoustical property is materially increased by mounting the fork on a resonance chamber, which reinforces the ground tone at the expense of the others, as will be explained in a later chapter.

Besides simplicity and purity of tone, the tuning-fork

---

[1] In consequence of the difference of rate in the transmission of sound through different media, the size of brass tuning-forks is much less than that of steel ones. "The velocity of sound in steel is at a maximum, amounting to 5,237 metres per second. For brass the velocity would be less in about the ratio 1·5 : 1. So that a tuning-fork made of brass would be about a fifth lower in pitch than if the material were steel."—Lord Rayleigh, *Theory of Sound* p. 220.

MODES OF PRODUCTION OF SOUND.

possesses another property which is extremely valuable for theoretical investigations; that namely of being only slightly

Fig. 11.—A tuning-fork mounted on a sounding-box.

affected by differences of temperature. Like all other metallic bodies, it expands according to a definite law, the coefficient of which expansion for each metal with given increments of heat is easily obtained; the modulus of elasticity of the material is diminished to a minute amount by the same cause. Hence all forks flatten somewhat with warmth. It is not very difficult, however, to apply a proper correction to this error, and they then become far the most trustworthy standards of pitch we are acquainted with.[1]

On the other hand their greatest disadvantage is due to the rapid falling off of vibration from internal friction and resistance of the air. Nor is it easy to find a means of exciting them which shall elicit a sustained tone. The usual excitant is a blow with a heavy body, covered with leather or flannel. French forks are made with a slight convergence between the inner sides of the prongs, and are excited by drawing a pin somewhat larger than the aperture through it from below. The writer has succeeded in keeping large forks of comparatively slow motion excited, by striking one prong gently but repeatedly with a small hammer attached to the mechanism of an electric "trembler" bell. But by far the best method is that largely and successfully employed by Helmholtz in the investigation of vowel sounds, and in his "Vibration Microscope," that namely of causing the fork itself, or one in harmonic relation with it, to become the contact breaker in a galvanic circuit, including an electro-magnet, which keeps up a series of synchronous impulses on the steel prongs of the tuning-fork itself. It has been common to place the electro-magnet on one side of the fork, but this is liable to draw down the fork into firm adhesion to its pole. Helmholtz places the two poles of a rather wide horseshoe-magnet outside the two prongs of the fork, so as simultaneously to draw them apart, but a still simpler and more effective method was illustrated in the Loan Collection by some French instruments, and is figured diagrammatically by Lord Rayleigh[2] in which a single short straight magnet, with wire wound in one coil around a core of bobbin shape is interposed between the two prongs, thus tending to draw them closer together at every contact without exerting any strain upon the supporting stem of the fork.

---

[1] Tuning-forks are among the instruments the use of which has extended from sound into other branches of physics, after a pleasant fashion of reciprocity. They have been employed as measurers of small intervals of time; their pendular vibrations are so regular, so accurate, and so easily adjusted to any one period of vibration, that they furnish an admirable means toward this end. A beautiful instrument of this nature was contributed to the Loan Exhibition at South Kensington by the French *Conservatoire des Arts et Métiers*.

[2] *Theory of Sound*, p. 56, where it is stated that such interruptors are within the capacity of a village blacksmith. The only drawback is the need of special wide forks; those usually made not leaving sufficient room for the interposed electro-magnet.

MODES OF PRODUCTION OF SOUND.

Summary of Vibrations in rods.

1. Transverse or lateral................................Harmonicons.
   - A. *Both ends fixed.* Resemble strings.
     Rise in pitch with thickness of rod.
     Depend on elasticity, not tension.
     Harmonics follow the squares of odd numbers.
   - B. *One end fixed.*..............................Musical Box.
     Vibration runs four times along the rod.
     Inversely proportional to square of length.
     Hence used by Chladni as Tonometers.
     Independent of area or cross section of rod.
     Harmonics follow the squares of odd numbers.
   - C. *Centre fixed.*................................Tuning forks.
     Pitch flatter than that of a straight rod.
     The middle nodal points approximate.
     First harmonic has four nodes.
     Harmonics generally distant from prime.
     Follow series of odd numbers. Often discordant.

Vibration of Plates.—When the rods or bars are expanded into plates, the second dimension of which bears a large ratio

Fig. 12.—Vibrations of a plate.

to their linear extension, we enter upon the study of the beautiful figures named after their discoverer Chladni, and subsequently investigated by Wheatstone.

26   ON SOUND.   [CHAP.

Chladni's method of investigation consisted in supporting plates of glass or metal, either square, circular, or of some other regular outline, by means of a kind of clamp, and bowing

Fig. 13.—Nodal lines of vibrating circular or polygonal plates, according to Chladni and Savart.

the edge in different points, with an ordinary rosined bow. The vibrations thus excited were analysed by means of sand previously strewn on the plate; it left the vibrating segments to heap itself on the quiescent nodes. A vast variety of beautiful figures was thus obtained, each corresponding to a particular note, and to a special mode of vibration. The more complicated forms were obtained by a combination of bowing and damping; the latter being accomplished by touching the edge of the plate in various places with the tips of the fingers, and thus hindering the motion of the spots touched. The rate of vibration in a disc was found to be proportional to its thickness and inversely proportional to the square of its diameter.

These experiments were continued by Faraday among others, who modified them by adding a light powder such as the spores of lycopodium mixed with the sand. This instead, like the sand, of seeking the nodal points of rest, collected at the places of most violent vibration, a phenomenon ultimately explained by the currents of air surrounding the vibrating plate, and not occurring *in vacuo*.

The laws which regulate the vibrations of plates have been much discussed by mathematicians; and indeed furnish problems of considerable complexity. But as no practical application of this method of eliciting sound occurs, it will be sufficient to refer the reader to the works in question for further details.[1]

Bells may be considered to hold the same relation to plates in their mode of vibration that the tuning-fork does to the linear bar or rod. Indeed gongs and some varieties of oriental bells are very little modified from the original shape of the plate.

The gong is usually a circular sheet of hard metal, rendered more elastic by hammering, strengthened at the edge by a deep flange of the same nature. The vibrations here are very complex and irregular, approximating somewhat to those of a tense membrane, and the note, if note it can be called, is a fortuitous combination of several discordant tones. Not very dissimilar from these is the cymbal; a noise-producing machine, intended, like the drum, rather to mark rhythm and accent than to emit a definite note. It differs from the gong in speaking from the treble register, and in intentionally reinforcing the high clashing harmonics. For this end the two cymbals are sharply struck together, touching in a few limited points; whereas the gong, like the string in the pianoforte, is gently but rapidly struck with a large soft beater over a

---

[1] Thomson and Tait's *Natural Philosophy*; Lord Rayleigh, *Theory of Sound*, 1. 293; Donkin; Kirchoff.

circle about one-third less in diameter than the instrument itself. No doubt this situation for the blows indicates the ring of greatest vibration discovered empirically. The same position, it will be found, produces **the best quality of tone** from the kettledrum.

Bells proper are usually in the form of vessels, either of hemispherical shape, as in clock-bells; or of a very complex outline, as in church and house bells. They give a note rich in **upper** partial tones, **or** harmonics, usually discordant **to the** foundation tone, as will be seen farther on, vibrating along the free edge, either in four, six, or more sections. The deepest note is given by the division of the edge into four segments, in which case the bell itself is momentarily disfigured by the blow of the hammer into an elliptical figure; returning by its elasticity into one with its major axis at right angles to that originally formed. Other

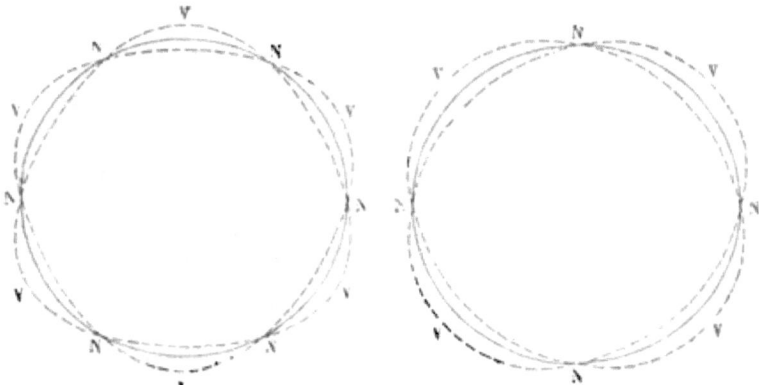

Fig. 14.—Nodes and segments of a vibrating bell.

segmentations of the vibrating **edge or** sound-bow can be produced, as in **the** plate, always **preserving an even** number of **4, 6, 8, 10, or** more vibrating **parts.** The sound of the hemispherical **bell is far** more pure and uniform than that of the ordinary **church bell.** Hence **this shape is** commonly used for **clock chimes, and** for carillons, **to which also its** compact outline, graduated **size,** and convenient **facility of stowage** in a limited space render it specially fitted. A large number of these hemispheres **can be** arranged on a single axis, with just so much of the free edge projecting **as** is needed for the hammer to strike **upon.** On the other hand, the clock-bell cannot be swung from gudgeons, and does not, from the shortness of its axis of figure, admit of an internal swinging clapper. It is therefore unfit for ringing in a peal.

The more elastic the material of which a bell is composed, the higher will be its **note.**[1] The necessary size and weight to elicit a given note are mainly matters of trial and experience, in consequence of which most of the older peals of bells are grossly out of tune. The common shape of church bells is that of a truncated conoid or paraboloid, closed at the apical end by a dome-shaped roof, to which the suspending luggs are attached. Assuming the diameter at the base as 15, and the height as 12, with a curvature below, in section, of radius 8 for the lower half, and for the upper of radius 30, Dr. Haughton has given a table of weights and pitch. One octave of these deserve quotation, in the hope that they may some day be constructed.

Note.	Diameter.	Sound-bow.	Weight.	Clapper.
	Inches.	Thickness.	lbs.	lbs.
CCCC	128·0	8·53	40,960	1,029
DDD	113·8	7·58	28,767	725
EEE	102·4	6·83	20,971	529
FFF	96·0	6·40	17,280	437
GGG	85·3	5·69	12,136	308
AAA	76·8	5·12	8,817	226
BBB	68·3	4·55	6,214	161

Helmholtz states that the tones vary with the greater or less thickness of the wall of the bell towards the margin, and that it appears to be an essential point in the art of casting bells to make the deeper proper tones mutually harmonic by

---

[1] Bells have of late been made of steel, but the common composition is of copper and tin, hence termed "bell-" or "gun-metal," in the approximate proportions of six atoms of the former to one of the latter. This is equivalent to a percentage by weight of copper 76·5, and tin 23·5. The exact atomic ratio seems to produce too hard and brittle an alloy, which defect can be reduced by slightly increasing the quantity of copper. A common mixture is 13 copper by weight to 4 tin, or by an admixture of zinc, and possibly of silver, although the traditionary stories as to the effect of the latter metal seem to a great extent imaginary.

giving a certain empirical form. Gleitz, the organist, in his "Historical Notes on the Great Bell and the other Bells in Erfurt Cathedral," states that the first-named gives the following series of notes : e, $g\sharp$, b, e', $g'\sharp$, b', $c''\sharp$. It was cast in 1477. Hemony of Zutphen, in the seventeenth century, required a good bell to have three octaves, two fifths, one major and one minor third. By the kindness of Sir F. G. Ouseley and Dr. Stainer, the writer is enabled to reproduce, in musical notation, the compound notes of several well-known bells. When played on the pianoforte the dampers should be raised.

HEREFORD CATHEDRAL.
{ Hour Bell.
{ Tenor Bell.

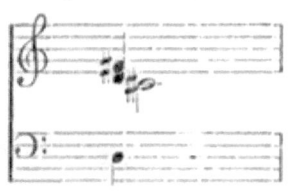

GREAT TOM OF OXFORD.

ST. PAUL'S.—Large Hour Bell.

BIG BEN OF WESTMINSTER.

(8ve lower)

In these notations, which can only be considered approximate, minims are employed for loud sounds and crotchets for those which are less prominent. All the notes should be struck simultaneously, but the minims louder than the crotchets.

In the large hour bell of St. Paul's, the upper note is much louder than the A flat, which ought to be the true note of the bell. The wavy line is to show that the pitch of the upper note does not remain stationary. In Big Ben the E is not a pure note, but is combined with so many sounds that it is impossible to give a nearer analysis of the tone of the bell.

It is found that the number of vibrations made in a given time is proportional to the square of the number of segments into which the bell divides itself. Thus, if $2m$ be the number of vibrating **segments**, we find, since n varies as m^2, by making m **successively** equal to 2, 3, 4, &c., for the notes of the bell.

$$\begin{array}{cccccc} 4 & 9 & 16 & 25 & 36 & \&c. \quad m^2 \\ 1 & \tfrac{9}{4} & 4 & \tfrac{25}{4} & 9 & \\ C_1 & D_2 & C_2 & G^\sharp_{\natural 3} & D_4 & \end{array}$$

corresponding to its division into the following segments:—

$$\begin{array}{cccccc} 4 & 6 & 8 & 10 & 12 & \&c. \quad 2m. \end{array}$$

None of these secondary notes are harmonics to C, except C_3. They are, however, very variously audible in different **bells**.

The figure of the usual church bell is irregular, more or less conical or cylindrical in outline, and only symmetrical around its vertical axis. The shape of bells varies much in different epochs and countries. They are **almost invariably made** of cast metal, therein differing from **the gong**. Near the open extremity is a thickened ring of metal termed the sound-bow, against which the clapper, ordinarily made of soft wrought iron, strikes. After a time the bell and clapper gradually mould themselves to one another, and the tone becomes fuller, with fewer of the very acute upper partials. But from accidental irregularities of shape, added to the varying specific gravity of cast metal, no church bell speaks with a single foundation sound, or with only a single **series of** harmonics. There is a common tendency to the after-production **of new** sounds, which encroach upon those originally emitted. Hence the weird effect of these grand sound-producers, the difficulty of fixing their exact note, and the mass of throbbing "beats" due to interference which can always be heard in the neighbourhood of a large bell. They are to a certain **extent** tunable by chipping away more or less metal at the sound-bow, **but the** main pitch has to be secured empirically through **the careful** imitation of a pattern known by experience to produce a certain note.

Bells may be **excited** in other ways than by striking. Small **specimens**, whether of glass or metal, answer well to the **fiddle-bow**, and their sound may also be developed by other forms **of friction**. In these cases the sound is sweet and continuous. This form of excitation has been brought to its highest point in the instrument known as the **Musical Glasses**, formerly in great repute, and recently revived. They were a collection of glass vessels, vibrating on the principle of bells, selected so as to form an approximate scale, and farther tuned by pouring in water. Their note was excited by rubbing the free edge either with the moistened finger or

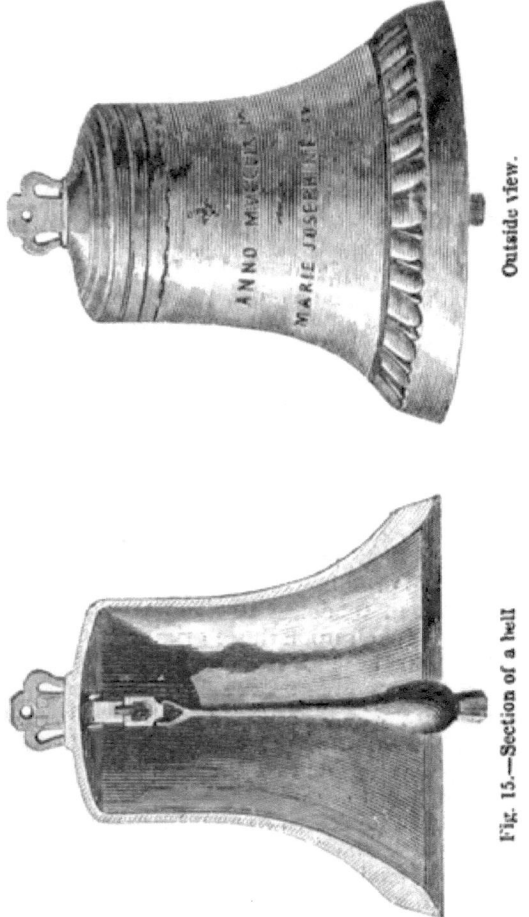

Fig. 15.—Section of a bell.

with a cloth wetted with acid. An ingenious modification of this arrangement exists at the South Kensington Museum, wherein the bells are fixed to a rotating spindle, and touched while in rotation with a wetted excitor. The sounds thus obtained are continuous, and of peculiarly luscious, though cloying, quality.

Lord Rayleigh[1] points out that though the pitch of the sound thus elicited is the same as that produced by a tap with the pad of the finger, the effect of the friction is in the first instance to cause tangential motion. By means of a neat

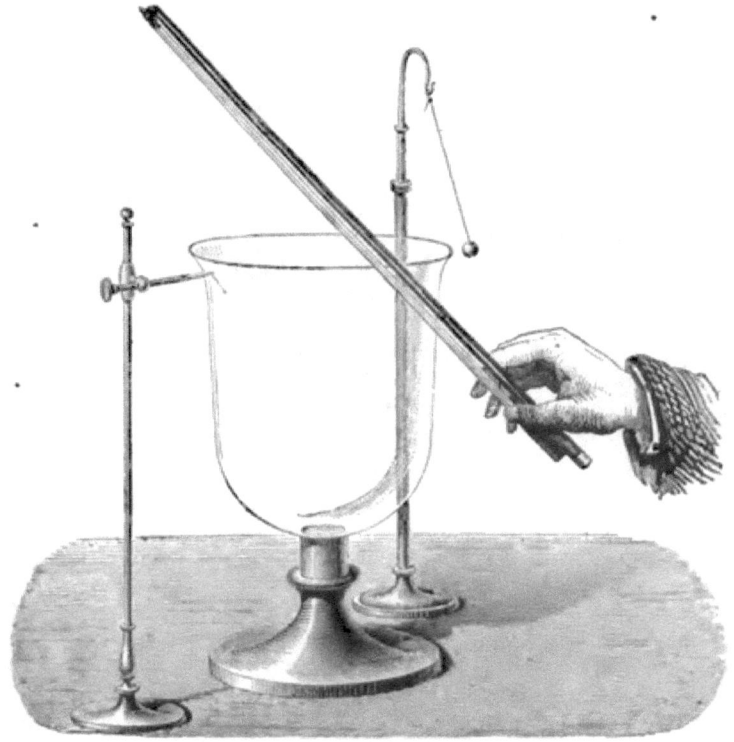

Fig. 16.—Proof of the vibration of a glass bell.

experiment with a large glass bell, a point on its edge was shown to revolve with angular velocity double that of the finger, the motion being alternately tangential or normal according to the relative positions of the finger and the point observed.

Membranes, as exciters of sound, occur chiefly in the form of drums, which stand in an intermediate position between mere noise-producers and sources of regular periodic vibration.

[1] *Opus citatum*, p. 325.

Their theory is in consequence somewhat complicated.[1] The theoretical membrane is a perfectly flexible and infinitely thin lamina of solid matter, of uniform material and thickness, which is stretched in all directions by a tension so great as to remain sensibly unaltered during vibration. Its vibrations have been investigated by Bourget in the same way as Chladni demonstrated those of plates. He excited them by means of organ-pipes, rendering the motion visible by means of sand scattered on the membrane. The principal results were:—

Summary of Vibrations in Membranes.

1. A circular membrane cannot vibrate in unison with every sound. It can only place itself in unison with sounds more acute than that heard when the membrane is tapped.

2. The sounds are separated by smaller intervals the higher they become.

3. Nodal lines are only formed distinctly in response to certain definite sounds. A little above or below, confusion ensues, and when the pitch of the pipe is decidedly altered the membrane remains unmoved.

4. The nodal lines are circles or diameters, and combinations of these. When the number of diameters exceeds two, the sand tends to heap itself confusedly towards the middle of the membrane.

As a matter of practice, membranes are but little utilized in music as sound-producers. Of the three kinds of drums, the side drum, the bass drum, and the kettle drum, only the last is tuned to a definite note, and that often very imperfectly. The side drum and bass drum are mostly used in military bands, and serve chiefly the purpose of marking rhythm and accent for marching. According, however, to the arrangement of modern orchestras, the kettle drum stands alone in possessing two, or at most three, notes of the sixteen-feet octave.

Zambomba.—A curious combination of a membrane with a rod vibrating longitudinally is used in Spain and Portugal under the above name. To the centre of a membrane like that of a tambourine is attached a small smooth rod of cane or strong straw. If this be gently rubbed between the wetted fingers the longitudinal vibrations thus excited in it are conveyed to the central point of the membrane, and a peculiar drum-like sound is produced, possessing a certain rhythm according to the rapidity with which the fingers are moved upwards and downwards.

Further remarks will be given on this subject in the chapter on special applications of acoustics to practical music.

Vibrations of Reeds.—Reeds form the natural link between solid and aeriform vibrators. Although they, as elastic tongues

---

[1] See Lord Rayleigh, *op. cit.* p. 290 for farther details.

of wood or metal, have their proper laws, following generally those of rods or bars fixed at one extremity and free at the other, they rarely if ever conform strictly to them, but are to a greater or less extent coerced by the elastic fluid which sets

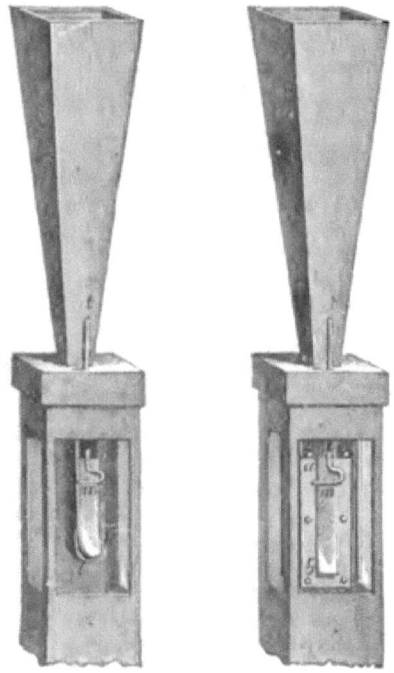

Fig. 17.—Striking Reed.   Fig. 18.—Free Reed.

them in motion, and from the consonance of which they mainly derive their power and quality of tone. When, as in the harmonium, they vibrate into the general mass of air, they are obviously less coerced than when they are adapted to cylindrical or conical pipes which possess their own note and vibration number. But even in the latter case, reeds, if of sufficient strength, are able to assert their independence; as is seen from the fact that the clarinet, a reed instrument of less than two feet in length, is able to reach the lowest notes of the four feet octave. This great extension of its compass can only be due to the coercive power of the slowly vibrating reed on the comparatively short column of air.

Reeds are usually divided into two kinds, the beating, and the free reed. The former is of considerable antiquity; the latter appears to be a modern invention. Both forms possess the elastic tongue; but the beating reed differs from the free in that the tongue is made to cover the orifice through which the air passes and slightly to overlap its edges; whereas in the latter it passes freely, as the name implies, though with but little room to spare, past the edges of the slit in which it is adapted. Orchestral reed instruments and all the older organ-reeds are of the beating kind.[1]

The beating reed is farther subdivided into single and double varieties, the former represented by the clarinet, and by the usual organ reeds; the latter appearing in the oboe and bassoon. These will be described in Chapter VIII.

It has often been pointed out that the type of all reeds is the little pipe of wheaten straw made by children in the fields. The hollow stem of a tall grass cut off just above a knot furnishes the stopped tube of the instrument, which is made to speak by slitting the straw upwards towards the knot for about an inch, thus leaving a tongue standing out and in a position to vibrate. The only instrument in which this primitive arrangement survives is the bagpipe, the "drones" of which, especially in old specimens, contain a large reed thus made from the woody tubular stem of some perennial plant.

Vibrations of Columns of Air.—Next in importance to strings as sound-generators may be ranked pipes and columns of air. A very simple experiment will illustrate this. If a piece of stout glass or metal tubing from a foot to two feet long, and an inch or more in diameter, be taken, its ends smoothed and rounded to a blunt edge, it will furnish the whole apparatus required. Holding it horizontally in one hand and striking the open end smartly with the palm of the other hand, sufficient vibration will be excited in the contained air to produce a distinct musical note, which often lasts a second or more; long enough, at any rate, for its pitch to be heard and determined. If, after striking, the hand be quickly removed, a second note is heard to follow the first at the interval of an octave above. In the former case the pipe vibrates as what is termed a stopped pipe, with one end closed, in the latter case as an open pipe. All the various forms of pipe used in the organ and elsewhere only differ from these rudimentary forms in

---

[1] In Dr. Tyndall's otherwise excellent and accurate work on *Sound*, to which the writer is much indebted, the organ reed is represented and described as a free reed. Free reeds have been found to fail in this position.

having a more complex mechanism for originating and maintaining the musical vibration.

When both ends of the tube are open, a pulse travelling backwards and forwards within it is completely restored to its original state after traversing *twice* the length of the tube, suffering in the process two reflections; but when one end is closed, a double passage is not sufficient to close the cycle of changes. The original state cannot be recovered until after two reflections from the open end, and the pulse travels over four times the length of the tube.

To make the unstopped tube in the above experiment yield the same note as the stopped, it would therefore be necessary to give it double the length. This law is universal, and may easily be explained.

But vibration may be set up in the column of air otherwise than by the blow above described. If a gentle stream of breath from the lips be sent obliquely across the open end of either an open or a stopped tube, an audible note results; indeed a common but simple instrument usually named the pandean pipes acts on this principle. A series of small reed tubes of graduated lengths are arranged side by side with their orifices upwards in a horizontal line, and their closed extremities downwards. They are fastened together by strips of wood. If the mouth be passed along the upper row of holes and the breath gently urged into them, a scale more or less correct according to the accuracy with which their relative lengths have been adjusted will result. Hence it is often termed the mouth organ. If a vibrating body such as a tuning-fork be brought opposite the orifice, and the length of tube be in a proper ratio to it, the note will immediately start out into prominence. This phenomenon will be further explained under the subject of consonance. In organ-pipes, flutes and flageolets, a thin sheet of compressed air is made to impinge against a sharp edge. This edge may be the sharpened extremity of the open tube itself, as is well seen in a remarkable example brought from Egypt by a friend of the writer's :—[1]

Fig. 19.—Egyptian flute.

We have here essentially an open pipe blown on the principle

[1] F. Girdlestone, Esq., of the Charterhouse, Godalming.

of the pandean closed pipe, but furnished with six lateral holes, by means of which the length of the vibrating column may be altered, and the note correspondingly raised. It affords an instructive link in the history of sound-production

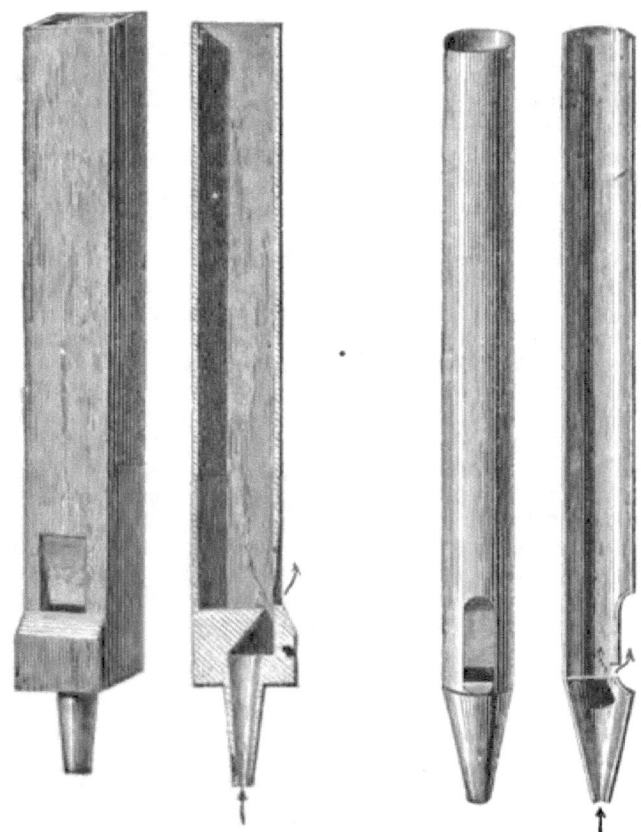

Fig. 20.—Prismatic sonorous pipes.   Fig. 21.—Cylindrical sonorous pipes.

which has not to the writer's knowledge been published before. In the ordinary flute, the upper orifice is closed with a cork, and the stream of air is passed over a hole with thinned edges perforated at the side near the top, the lower series of holes remaining unaltered.

In the organ pipe, a more complicated arrangement is present which is best seen by means of **figures.**

From the wind-chest a tube leads into a cavity the only outlet of which is a linear crack forming the **foot** of the pipe. **Just over this** fissure the wood or metal is cut away so as to **leave a** feather-edged portion, communicating with the interior of the pipe, exactly splitting the stream of wind. An explanation has of late been tendered as to the action here set up. The flat plate of compressed air blown through the slit is compared to the elastic material of a vibrating reed. In passing across the orifice it momentarily produces a slight exhaustive or suctional effect, tending to rarefy the air contained in the lower part of the pipe. This by the elasticity of the air soon sets up a corresponding compression, and **the** two allied states react upon the original lamina of air issuing **from** the bellows, causing it to vibrate and to communicate its motion to that within the pipe. Indeed, the course of the air currents can to a certain extent be demonstrated by feeding the pipe with tobacco smoke or some other semi-opaque vapour.

Motion of Air in Pipes.—Schneebeli drove air rendered opaque by smoke through a moveable slit. When it passed entirely outside the pipe no sound was produced, but appeared when the issuing sheet was gently blown on at right angles, continuing when once started until a counter-current was produced by blowing down the upper orifice of the pipe. Little **or no** smoke penetrated into the pipe. **If** the sheet of air passed into the pipe entirely there was also no sound, but on blowing **into** the upper end sound was produced. He concludes that **the** *Lüft-Lamelle*, or aerial lamina, acts a part analogous **to that** of the reed in reed-pipes. Hermann Smith has come **by** independent observations to a similar conclusion, terming **the** sheet of air an "aeroplastic reed." Schneebeli considers its effect to be condensatory; Hermann Smith, with far greater probability, holds it to be exhaustive, and similar to the common spray-producing apparatus. The tones **of** the air reed **and** pipe he believes to be distinct, that of the former being **far more** acute than the latter, and sometimes capable of coercing **it.** There can be no doubt that this condition of things exists in the case of the reed fitted with **a consonant** pipe, as in the case of the clarinet given above.

In all the **above** cases, the air, like the string or the **rod,** may assume several modes of undulation. In the **open pipe,** the embouchure **at** which the wind enters is obviously a place of greatest motion, corresponding to the ventral segment of **a** string. So also will be the upper open extremity. Half way between these, at the point where the two opposite and correlative motions meet and neutralize one another, will be a node or point of rest. In this instance the pipe will give its lowest or fundamental note. If the force of the current be increased, a shorter wave may be set in action, a node being established at one-fourth of the whole length from the embouchure, and another at the same distance from the top. The pipe then speaks its first harmonic, the octave of the fundamental. By a further wind-pressure three nodes may form, the first of which is one-sixth from the embouchure, the third at a similar distance from the top, and the second halfway between the other two; the pipe giving its second harmonic a twelfth above the foundation. As the lengths of the waves are in the proportion $\frac{1}{2}, \frac{1}{4}, \frac{1}{6}$, it is obvious the times of vibration will be 3, 2, 1, or corresponding to the series of natural numbers.

Midway between each consecutive pair of nodes there is a *loop*, or place of no pressure-variation. At any of these loops a communication may be established with the external air, without causing any disturbance of the motion. The loops are places of maximum velocity, and the nodes those of maximum pressure-variation.

In stopped pipes a different law obtains; for the waves have clearly to traverse twice instead of once the length of the tube, being returned by the closed extremity. This fact also influences the position of the nodes. When the fundamental note is struck the column is unbroken, the only node being at the stopped end. In sounding the first harmonic another node is set up at one-third of length from the open end. With the second harmonic, a node forms at one-fifth of length from the open end, the second dividing the lower four-fifths into two equal parts. In any case, the stopped end must be a node, so that the second form of vibration of the open pipe, and all others which would render it the centre of a ventral segment, are excluded. Hence the harmonics of a stopped pipe follow the series of the odd numbers 1, 3, 5, &c. These relations between the fundamental note of a tube and its overtones were discovered by Daniel Bernoulli and are generally known as the **Laws of Bernoulli**. When the length of a tube exceeds its diameter considerably, the note is independent of the latter, and varies with the length alone. In both stopped and open pipes, the distance from an open end to the nearest node is a quarter-wave length of the note emitted. In the open pipe there is no further limitation; but in the case of the stopped pipe, the nearest node to the mouthpiece must also be distant an even number of quarter wavelengths from the stopped end, which is itself a node.

These distinctions hold good with pipes of which the bore is cylindrical or prismatic with parallel sides. It was, however, shown by Wheatstone that a pipe of conical bore, while giving out a similar fundamental note to one of the same length of cylindrical shape, differs as regards the position of the nodes when emitting one of its harmonics. The first node, for instance, in an open conical pipe is not in the middle, but some way towards the smaller end. In conical stopped pipes, therefore, or in instruments which resemble them, such as the oboe and bassoon, the even harmonics are not necessarily excluded. In the clarinet alone, of which the bore is truly cylindrical, they are not to be detected.

It appears from more modern observations that the Laws of Bernoulli require a correction which will be given in a later chapter.

Vibrations from Heat.

Trevelyan's Rocker.

A heated brass or copper bar so shaped as to rock readily from one point of support to another, is laid upon a cold

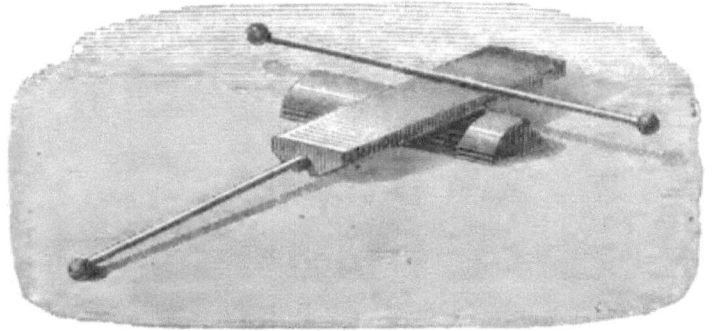

Fig. 22.—Trevelyan's instrument.

block of lead. The communication of heat through the point of support expands the lead lying immediately below in such a manner that the rocker receives a small impulse. During the interruption of the contact, the communicated heat has time to disperse itself in some degree into the mass of lead, and it is not difficult to see that the impulse is of a kind to encourage the motion, and to produce sound.

Fig. 23.—Trevelyan's instrument. Cause of vibratory movements.

Sondhauss's Experiment.

When a bulb about ¾ of an inch in diameter is blown at the end of a narrow tube 5 or 6 inches in length, a sound is sometimes heard proceeding from the heated glass. It was proved by Sondhauss that vibration of the glass is no essential part of the phenomenon. An explanation of the production of sound has been given by Lord Rayleigh, which will be fully detailed in Chapter VIII.

Singing Flames.—Although the series of small explosions by which the combustion of gas is marked have contributed some brilliant experiments to physicists, they can hardly as yet be said to be a practical source of musical sound. The flame of hydrogen has long been known as a means of originating a note; and of late a second form of the experiment has utilized the ordinary flame of coal-gas as a very delicate consonator and test for sounds produced extraneously in its neighbourhood. The former are termed singing, the latter sensitive, flames. The sensitive flames may best be considered elsewhere; but a short account of the singing-flame is required to complete the series of sound-producers.

It is easy, whenever a jet of hydrogen is inflamed, or even when coal-gas issues from a burner with some force, to hear an unmusical hissing or roaring accompanying the process. Even in this case, the noise often puts on a more or less definite form of vibration, and an impure note of coarse quality makes itself manifest. But if the jet be surrounded by a resonating tube, this has the power of reducing the irregular vibrations to a greater uniformity, and of selecting those which synchronise with its own vibration-period. The note in this case is often very pure and powerful. The general type of the process may be studied in the ordinary Bunsen burner. In this very convenient laboratory appliance. ordinary coal-gas is allowed to issue by a small orifice into a larger tube, perforated at its lower extremity with several holes admitting more or less atmospheric air, the admission of which is regulated by a slide. An explosive mixture of gas and air is thus made, which is prevented from communicating with the source of gas by the cooling effect of the surrounding tube, just as occurs in the wire-gauze envelope of a Davy safety-lamp. If only a small amount of

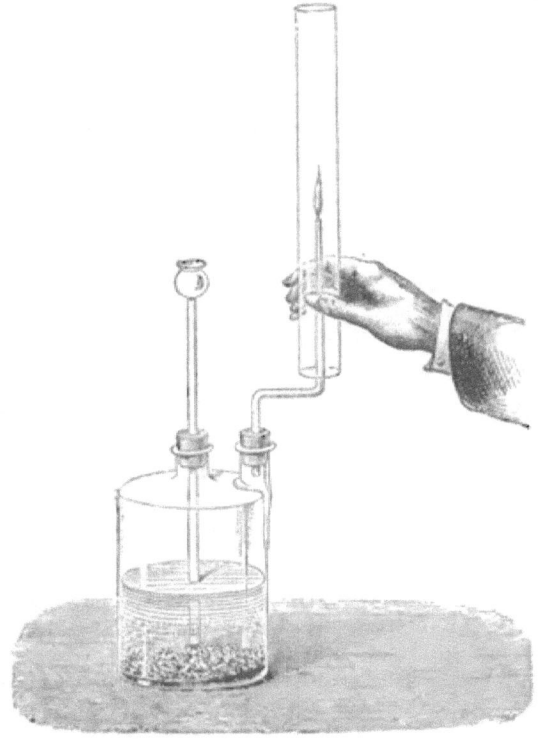

Fig. 24.—Philosophical lamp or chemical harmonicon.

air is admitted the mixture burns with a semi-luminous flame and silently. But as the proportion of air is allowed to increase by opening the slider, the flame loses its luminosity and at length begins to roar. The combustion gradually becomes discontinuous, and is indeed composed of a series of short successive explosions. At length the mixture becomes too explosive even for this, it lights throughout its whole length with emission of a single report, and carries the combustion down the tube to the small gas-jet itself. If the Bunsen burner, freely supplied with air, be introduced into a tall vertical tube, the roaring is toned to consonance and gives a pure note instead of an irregular roar. It is not always easy with quietly burning coal-gas in a tube to obtain any sound at all. But if the flame be reduced in size, and moved up and down the consonating tube, it may be observed to become tremulous at certain spots, and if left there, suddenly, by increase of the pulsations, bursts into sound. This action may be determined by the method first named as a means of exciting musical oscillation in a pipe, namely, by striking a slight blow with the palm of the hand on the upper orifice of the tube. The wave thus sent downwards is instantly taken up by the flame, and the note starts out, sometimes with such vehemence as to extinguish it altogether. It can also be set going by the voice. Many of the older works on chemistry and physics give the simple experiment with a hydrogen flame and a glass tube. Dr. Higgins names it in 1777; but Wheatstone was the first who attempted successfully to produce a definite scale of notes by this method. His instrument is now preserved in the Museum of King's College, and was recently shown at the Loan Exhibition at South Kensington. A series of glass tubes, with metal sliders for the purpose of tuning, are arranged in a row like the pipes of an organ. Within each of these is a fine conical tube pierced above with a capillary orifice. The lower end slides air-tight in a second tube connected with a supply of gas. In front is a short keyboard like that of a piano. Each key, on being depressed, lifts the small gas-jet from its position at the bottom of the tube to about the junction of the lower and two upper thirds, which, being a sensitive point, immediately originates the fundamental note of the tube.

The Pyrophone.—M. Kastner has endeavoured to utilize this principle in a musical instrument, but on a slightly different system. He says, "If two flames of a certain size be introduced into a tube made of glass, and if they be so disposed that they reach the third part of the tube's height, measured from the bottom, the flames will vibrate in unison. The phenomenon continues as long as the flames remain apart, but as soon as they are united, the sound ceases." By means of finger-keys the flames are united and separated so that a melody can be played. There is some uncertainty about the instrument, depending on the pressure of gas; so that although announced for performance in Paris, it has not hitherto been used.

A curious accidental source of sound appears to have been several times discovered. It possesses no practical importance, but affords an apt illustration of the theory of harmonics.

If a piece of the ordinary vulcanite tubing, such as is used for conveying gas, and which is prevented from collapsing by a spiral wire coiled round its internal surface, be cut to a length of about 18 inches, and gently blown into, a soft musical note of feeble but reedy quality is produced. On pressing the force of wind, it rises successively to higher notes, which will be found on examination to follow roughly the order of the common chord to the foundation tone. It is obvious that the wire coiled inside the tube produces a series of equidistant obstacles, competent to throw the air into regular vibration; the rapidity of which vibration, and the consequent pitch of the note produced, vary with the speed at which the air is blown into the calibre of the tube.

The only remaining source of sound is the human voice; but this is of so much importance that it will be considered separately in a later chapter.

CHAPTER II.

MODES OF PROPAGATION **OF** SOUND. **VELOCITY.** WAVE-MOTION. REFLECTION. **REFRACTION.**

The Propagation of Sound appears to take place to some extent through all bodies, but in **very** different amounts and with varying degrees of velocity. This factor has been found to vary directly as the square root of the bodies' elasticity, and inversely as the root of its **density**. The formula

$$V = \sqrt{\frac{E}{D}}$$

therefore serves for all **forms of matter.** Solids, **however,** being liable to many kinds **of strain,** and fluids, **whether liquids or gases,** to only one, **we may have** different values of E, and different velocities of **transmission** for the same solid. In a perfectly free solid this value of E is identical with **Young's modulus.** The **great majority** of solids however transmit sound more rapidly in one direction than in others. In solids, moreover, the thermal correction, to be **spoken of** presently, **is very small,** as it is also in fluids, whereas **in air** it is large.

By the Earth.—There **is distinct** evidence of its transmission through the **solid mass of** the earth itself for long **distances.** Humboldt says, "At Caracas, in the plains of **Calabozo,** and on the borders **of Rio Apure,** one **of the affluents of** the Oronoko; that **is to say,** over an extent of **130,000 square** kilometres, one hears a frightful **report, without experiencing** any shock, **at the** moment when a **torrent of lava flows from** the volcano St. Vincent, situated in the Antilles, **at a distance** of 1,200 kilometres. At the time of the **great eruption** of Cotopaxi in 1744, the subterranean reports **were heard** at Honda, on **the** borders of Magdalena; yet the **distance between** these **two points is** 810 kilometres;

their difference of level is 5,500 metres, and they are separated by the colossal mountainous masses of Quito Pasto and Popayan, and by numberless ravines and valleys. The sound was evidently not transmitted by the air but by the earth, and at a great depth. At the time of the earthquake of New Granada in February, 1835, the same phenomena were reproduced in Popayan, at Bogota, at Santa Maria, and in the Caracas, where the noise continued for seven hours without shocks; also at Haiti in Jamaica, and on the borders of Nicaragua."[1]

No better illustration of the conveyance of sound by solid media can be found, than that which occurred in the recent colliery accident (1877). Coal is an excellent conductor of sound, being both light and elastic. It was possible from the very first of the noble attempts to rescue the five imprisoned miners, to communicate with them through a long barrier of intervening coal, by knocking on the external surface of the seam in which they were incarcerated. In the same way they were able, as it were, to telegraph back the fact of their existence to their rescuers.

Mons. Biot experimented on a cast-iron pipe 951 metres in length, and found that sound is propagated through this metal with a mean velocity of 3,250 metres a second, or more than $9\frac{1}{2}$ times that through air of the same temperature. The pipe used was of rather heterogeneous material, a fact which renders the quantitative determination somewhat doubtful.

An ingenious application of the principle of propagation through solids occurs in Wheatstone's *Telephone*, exhibited at the Polytechnic Institution many years ago. A band of performers, with violin, clarinet, piano, and other instruments, were placed in a basement room, through the ceiling of which rods of fir-wood were passed into a concert-room above. Each rod was attached by its lower extremity to one of the instruments, at its upper end it was connected with a consonator such as will be described later. When the instruments were played, not only the actual notes, but even the quality and character of each were distinctly audible to any number of listeners in the concert-room. It will be seen from the Table that fir-wood transmits sound at the enormous velocity of 5,994 metres per second, or more than eighteen times that of its transmission in air.

A clever toy has been lately sold by which the transmission of sound through solids may be simply demonstrated. It consists of two tin cylinders, each closed at one end, and

---

[1] Quoted in Guillemin's *Forces of Nature*.

joined together by means of a wire, or even an elastic string, of several yards length. A sentence gently spoken into one cylinder can be distinctly heard by applying the ear to the orifice of the other, when it is quite inaudible from distance through the air.

The Telephone of Graham Bell acts on a totally different principle, converting the vibrations of a metallic plate into magneto-electric currents in a coil of wire surrounding a small magnet. By an exactly similar apparatus at the other end of the conducting line, the undulatory currents thus produced are reconverted into musical tones.

The Microphone of Prof. Hughes really substitutes for a feeble sound, one much louder produced by varying resistance between opposed conductors. It is therefore essentially a relay.

Velocity in Air.—The velocity of sound in air has been the subject of many experiments since the time of Newton. Those of Goldingham, published in the *Philosophical Transactions* for 1823, of Arago in 1825, of Myrbach and Stampfer at Vienna, of Moll and Van Beek in Holland, and of Gregory, seem the most trustworthy. The observations have generally been those of the flash and the report of a distant cannon. The same observer notes both phenomena with the same watch, and if the distance of the gun be several miles, there is ample time to write down the observation of the flash, before preparing for observation of the sound.

Table of Experimental Determinations of Sound-Velocity.

	Metres at $0°$ cent.
1. Academie des Sciences, 1738.	332
2. Benzenberg, 1811 (mean).	333
3. Goldingham, 1821	331·1
4. Bureau des Longitudes, 1822	330·6
5. Moll and Van Beek	332·2
6. Stampfer and Myrbach	332·4
7. Bravais and Martin, 1844	332·4
8. Wertheim	331·6
9. Stone, 1871	332·4
10. Le Roux	330·7
11. Regnault	330·7

It has, however, been pointed out by Airy[1] that there is a physiological circumstance, the effects of which have hitherto escaped notice, but which probably produces a sensible error; it is that two different senses, sight and hearing, are employed

---

[1] *On Sound and Atmospheric Vibrations, with the Mathematical Elements Music.*

II.]   MODES OF PROPAGATION OF SOUND.   49

in noting the two phenomena, and we are not certain that impressions are received by them with equal speed. Indeed we believe that the perception of sound is slower by a measurable quantity, perhaps ·02″, than the perception of light, and this may affect the result with an error amounting to some hundreds of feet. It would be preferable if two observers noted, in the same manner, the time of the sound passing two isolated points. By using signals given reciprocally from two stations beyond both the observing points it will be easy to obtain a result for the time of passage of the sound, independent of the habits of each observer, independent of the different indications of their timekeepers, and independent of the velocity of the wind.

It is possible that a still closer determination might be made by adding to the Astronomer Royal's excellent method some form of electric chronograph, and perhaps the recording phonautograph described later on.

In Gases.—The *velocity of sound in gases is directly proportional to the square root of their elasticity, and inversely proportional to the square root of their respective densities.* The most remarkable case is that of hydrogen, which being about sixteen times lighter than oxygen, conveys sound about four times as fast.

The velocity being a function of the elasticity and density of the medium conveying the sound, the variation of either factor will cause it to be more or less rapidly propagated. Air in a close vessel, unable to expand, when subjected to heat, transmits sound more rapidly than when cooler. Air, moreover, expanding freely with heat, becomes rarefied, and this diminution of density, with unaltered elasticity, has a similar effect.

At a freezing temperature, the velocity has been found to be 1090 feet in a second.

The density of hydrogen being much less than that of air, and the elasticity the same, the fact above stated is fully accounted for. The reverse is true of carbonic acid, a very heavy gas.

The relation of the two is best expressed by the simple mathematical formula above given.

$$V = \sqrt{\frac{E}{D}}$$

V being the velocity, E the elasticity, and D the density. The law of Boyle and Marriotte, "that the temperature being the same, the volume of a mass of air is inversely as the pressure it supports," shows that this is true of all gases within certain limits. Hence the velocity, on high mountains, or even at the bottom of mines, does not vary if the temperature is constant. With an increase of temperature sound travels faster, with a decrease, slower. The rate of transmission is increased about two feet for each degree Centigrade, or 1·14 feet for each degree Fahrenheit. At the temperature of 60 degrees Fahrenheit we may reckon the velocity of sound at about 1,120 feet per second, or $12\frac{1}{2}$ miles per minute.

Velocity in		
Carbonic Acid	262	metres per second.
Oxygen	317	,, ,,
Air	331	,, ,,
Hydrogen	1270	,, ,,

By this means the distance of a sonorous body may be roughly measured, the velocity being about a mile in $4\frac{5}{8}$ seconds at medium temperatures.

The depth of the well in Carisbrook Castle has thus been determined, by dropping in a stone, and watching for the sound in the water, allowing of course a correction for the time of fall. The clock of the Houses of Parliament strikes the first blow of the hour within a second of Greenwich time; but five or six seconds have to be allowed for transmission of the sound to even moderately distant stations.

Velocity in Liquids.—The velocity with which sound is propagated in liquids was admirably demonstrated by the classical experiments of Colladon and Sturm, made in the Lake of Geneva. The observers were stationed in two boats on opposite sides of the lake. The sound was emitted from a bell struck by a hammer under the water, and received by a long speaking-tube with a vibrating plate covering its larger orifice, which was sunk vertically in the lake at the other station, the ear of the listener being applied to its smaller end. At the moment of striking the bell some powder was lighted by a match fixed to the hammer, and the determination was made by counting with a chronometer the time elapsing between the flash and the sound. The stations were determined to be 13,487 metres apart; the interval was $9\frac{1}{4}$ seconds; thus giving 1,435 metres for the velocity per second in water at 8° Centigrade. It appears, from subsequent experiments, that temperature causes considerable variation in the rate of transmission even in fluids, though far less than in gases; the velocity in the Seine at 15° Centigrade having been 1,437 **metres, in** sea-water at 20° – 1,453, and **at** 23° – 1,460.

Ether at 0° centig.	1159 metres per second.
Fresh Water at 15° centig.	1437 ,, ,,
Sea Water at 20° centig.	1453 ,, ,,
Ditto at 23° centig.	1460 ,, ,,

Velocity in Solid Bodies.—It is owing to the high elasticity of solid bodies such as glass and steel, that the velocity of sound-transmission in them is so great, in spite of their increased density.

The simplest mode of demonstrating this velocity is by means of the Sonometer for longitudinal vibrations of wires already named.

If two wires be stretched side by side in this apparatus, of equal length and thickness, but of different material, the notes, which have been already stated to be independent of tension, will be found to differ considerably. For instance, if they be of steel and brass, the former will be the sharper, owing to the greater velocity in the more elastic metal. With iron and brass the ratio is that of 11 : 17, representing an approximate velocity of 11,000 feet per second in the latter, and of 17,000 in the former.

Other methods of determination are given further on.

Some of the principal determinations may here be summarized.

Tin	2498 metres per second.
Silver	2684 ,, ,,
Platinum	2701 ,, ,,
Oak, Walnut	3440 ,, ,,
Copper	3716 ,, ,,
Steel and Iron	5030 ,, ,,
Glass	5438 ,, ,,
Fir Wood	5994 ,, ,,

Table of Sound-Propagation.

Occurs in all elastic bodies—

I. In Solids.

    1. The most rapid of all forms of matter.

    2. In free solids $V =$ Young's modulus of elasticity.

    3. The changes from heat very small.

    4. Solids not isotropic.

Computational determination follows from $V = \sqrt{\frac{E}{D}}$, which **applies to** matter generally.

Experimental—
1. **Strings**, longitudinal vibrations.
2. Kundt's experiments.
3. Wertheim's experiments.
4. **Biot's** measurement in cast-iron.

II. In Liquids.
Computational—

In water $V = \sqrt{\dfrac{E}{D}} = \sqrt{\dfrac{1033 \times 931}{\cdot 0000457}}$

$= 1489\cdot2$ metres per second.

Experimental—
1. Colladon and Sturm.
2. Wertheim's experiments.

III. In Air and **Gases.**
Computational—
1. **Newton,** *Principia.* ⎱ See below.
2. **Laplace's** correction. ⎰

Experimental—
1. Table **above given.**
2. Kundt's experiments. ⎱ See below.
3. Bosscha's experiments. ⎰

Wave Motion.—It must, however, be clearly understood that the **velocity thus** spoken of does not **imply the** translation **of material** particles from **one terminus to the other.** There **is nothing** resembling the **flight of a rifle** bullet **between a source of** sound **and the observer's ear.** The process is **essentially one of** wave motion ; a condition in which, though **each individual** particle **passes through a** very small distance from its original position **of rest, it** propagates the imparted impulse to its neighbours, **and each** neighbouring particle to those successively in contact with it.

It is therefore **necessary here to** advert briefly **to wave** motion generally, and to **the theory** of undulation **as applied to** sound.

"**The** theory **of the** transmission of sound through the **air,"** says Professor Airy,[1] "**as well as** through other bodies, **is** especially founded upon the conception of the transmission **of** waves, in which the nature of the motion is such that the movement of every particle is limited, while the law of relative movement of neighbouring particles is transmitted to **an** unlimited distance, either without change, or with change

---

[1] Airy on *Sound and* Atmospheric *Vibrations.*

following a definite law. In sound we have states of condensation and states of rarefaction, travelling on continually without limit in one direction; while the motion of every individual particle is extremely small, and is alternately backwards and forwards. This is the conception of a wave as depending on the motion of particles in the same line as that in which the wave travels. But there are other kinds of movement of particles, which are equally included under the conception of wave. The motion of the particles may be entirely transverse to the horizontal line; here it is not states of condensation and rarefaction that travel continually in the same direction, but states of elevation and depression that so travel. This is the kind of wave which is recognised as applying to Polarized Light. But in all these there is one general character; that a state of displacement travels on continually in one direction without limit; while the motion of each particle is, or may be, small and of oscillatory character. This is the general conception of a wave. The idea appears to have been first entertained by Newton, and was certainly first developed by him, for the purpose of explaining, what till then was totally obscure, the transmission of sound through air; it is worked out in the third book of the *Principia*, and among the many wonderful novelties of that wonderful work, it is not the least interesting or the least important. The mere conception of the motion of particles in the way pointed out is a very small part of Newton's work; the really important step is to show that the condensations and rarefactions produced by these motions will, by virtue of the known properties of air, produce such mechanical pressures upon every separate particle, that the different changes of motion which those pressures will produce on each individual particle, will be such that the assumed laws of movement will necessarily be maintained."

To reconcile his theoretical inference for the velocity of sound with observed measurements, he suggested the idea that the dimensions of air-particles produced a sensible effect. It has been subsequently discovered that the apparent discrepancy depends on changes of temperature developed by the ultimate rarefactions and condensations of which the sound wave consists.

Newton[1] originally computed the velocity at 0° Centigrade to be 916 feet per second. He took into account only the change of elasticity resulting from a change of density,

---

[1] Lee's *Acoustics, Light, and Heat.*

but entirely overlooked the augmentation of elasticity resulting from a change of temperature. Laplace was the first to show the true cause of this discrepancy, and applied a correction to Newton's investigation which brings the theoretical into complete accordance with the observed velocity. This consists in multiplying Newton's velocity by the square root of the ratio of the specific heat of air at constant pressure (C^p) to its specific heat at constant volume (C^v). Thus, if V be the calculated velocity, and V' the true velocity, then

$$V' = V \sqrt{\frac{C^p}{C^v}}$$

The value of the ratio $\frac{C^p}{C^v}$ is 1,414. Hence we have true velocity $= 916 \times \sqrt{1\cdot 414} = 1090\cdot 04$.

Wertheim's Experiments.—If V be the velocity of sound in a particular gas, in feet per second; λ the wave-length of a given note in this gas, and n the vibration number, then λ is the distance travelled in $\frac{1}{n}$ of a second; that travelled in a second $V = n\lambda$. - n being constant for all media, and V varying directly as λ, the velocities in two gases may be compared by observing the lengths of columns which give the same note. In columns of equal length, the velocities are directly as the vibration-numbers of the notes emitted. The results in some gases are given above.

Similar determinations were made by him by this method, in liquids and solids. The velocity in ether and alcohol was found to be 1,160 metres per second, in solution of chloride of calcium 1,900 metres.

Bosscha's Method.—This method depends on the precision with which the ear is able to decide whether short sharp sounds are simultaneous or not. Two small electro-magnetic counters are controlled by an interrupting fork, whose period is $\frac{1}{15}$ second, giving synchronous ticks. As one counter is gradually removed from the ear, the two series of ticks fall asunder. When the distance is 34 metres, coincidence again takes place, that being the distance travelled by sound in $\frac{1}{15}$ of a second.

Wave-length is the distance which sound travels in any medium during the period corresponding to the note sounded. Now the period which is the reciprocal of the vibration number, or frequency, is longer the lower the pitch of the note, and is a measure of that pitch.

Wave-length may therefore be taken as a measure of pitch if the medium in which the sound travels be known; but it varies in passing from one medium to another. It varies also according to the temperature, being lengthened when this rises, owing to the decrease of density thus produced.

On the other hand, the measurement of wave-length is of importance in determining the size of cavities used as **resonators**. For instance, the ordinary A of the French normal diapason makes 435 double vibrations in a second. Taking the velocity of sound in air at $50°$, as 1110 feet per second, the wave-length $= \frac{1110}{435} = 2$ ft. 6·602 in.; a resonance box having a quarter of this wave-length or 7·65 in. will be found to produce the most complete consonance. The sharp A of 456 vibrations per second used in English orchestras has a wave-length of only two ft. 5·2 in. giving 7·3 in. for its resonance box.

Kundt's determination of the Velocity of Sound is susceptible of considerable accuracy, and requires only simple apparatus. A glass tube about two metres long and two inches in diameter is closed at one end by a stopper, the other being fitted with a cork perforated by a smaller rod of glass or other elastic material bearing at its inner extremity a piston fitting smoothly into the larger tube. By rubbing the projecting end of the rod it is set into longitudinal vibration which is communicated to the air in the section of the tube between the stopper and the piston. Some light powder such as the spores of Lycopodium is contained in the tube; they are set into active vibration arranging themselves in small linear patches or heaps representing the nodes. The mean distance between these is equal to half a wave-length in air. If the rod be grasped at its middle by the cork, the wave-length of the sound it emits is twice its length. As the velocity of sound in any body is equal to the wave-length in that body multiplied by the number of vibrations in a second, if this latter be the same in both cases, the velocity of sound in the rod is to that in air as the length of the rod is to the distance between the heaps. If, for instance, the rod be of glass, and clamped in its middle, and the distance between the cork and piston be of the same length, the number of heaps will be found to be 8, corresponding to a velocity of sound in glass 16 times that in air. The method can also be employed for measuring velocity in other gases than air by introducing them into the tube and comparing the distances of the heaps with the lengths of the vibrating rod. By varying the material of the rod, the velocities of sound in various elastic solids may be also obtained. Brass was thus found to give a velocity of 10·87, steel 15·24, carbonic acid 0·8, coal-gas 1·6, and hydrogen 3·56, air being 1·0, or nearly as the inverse squares of their respective densities.

In a double apparatus devised by the same experimenter, the sounding tube was caused to vibrate *in its second mode* by friction applied near the middle, and thus nodes were formed at points distant from the ends by one-fourth of the length of the tube. At each of these points connection was made with an independent wave-tube, provided with an adjustable stopper, and with branch tubes and stop-cocks suitable for admitting various gases to be experimented on. The dust-figures in either tube thus correspond rigorously to the same pitch, and therefore comparison of the intervals of their recurrence gives a correct determination of the velocity of propagation for the two gases with which the tubes are filled.

A few of the results arrived at were—

1. Velocity of sound in a tube diminishes with diameter, but above a certain diameter the change is not perceptible.

2. Diminution of velocity increases with the wave-length.

3. The presence of powder, or roughening the interior of the tube, diminishes the velocity, especially in small tubes.

4. In wide tubes the velocity is independent of pressure, but in small tubes increases with it.

5. All changes of velocity are due to friction, and to exchange of heat between the air and the sides of the tube.

6. The velocity of sound at 100° agrees exactly with that given by theory.

Fig. 25.—Kundt's tube.

Reflection and Refraction of Sound.—When sound-waves meet a fixed obstacle, they are reflected, just as occurs in the case of light. The two sets are propagated as if starting from separate sources. In the case of a flat surface the reflected undulations seem to diverge from a source situated behind it, which corresponds to the virtual image of a plane mirror. The angles of incidence and reflection, as in the case of light, are equal.

MODES OF PROPAGATION OF SOUND.

By means of concave mirrors sound can easily be collected or concentrated to a focus, as may be seen in the illustration subjoined.

Fig. 26.—Experimental study of the laws of the reflection of sound.

It can also be refracted by lenses made of collodion, and filled with a dense gas, such as carbonic acid. The figure given on the next page is that of M. Sondhauss's instrument.

58 ON SOUND. [CHAP.

Echoes.—The only important practical illustration of these physical facts occurs in the case of echoes. It appears that the sensation of sound occupies about the tenth of a second, during which time the sound-wave travels about 34

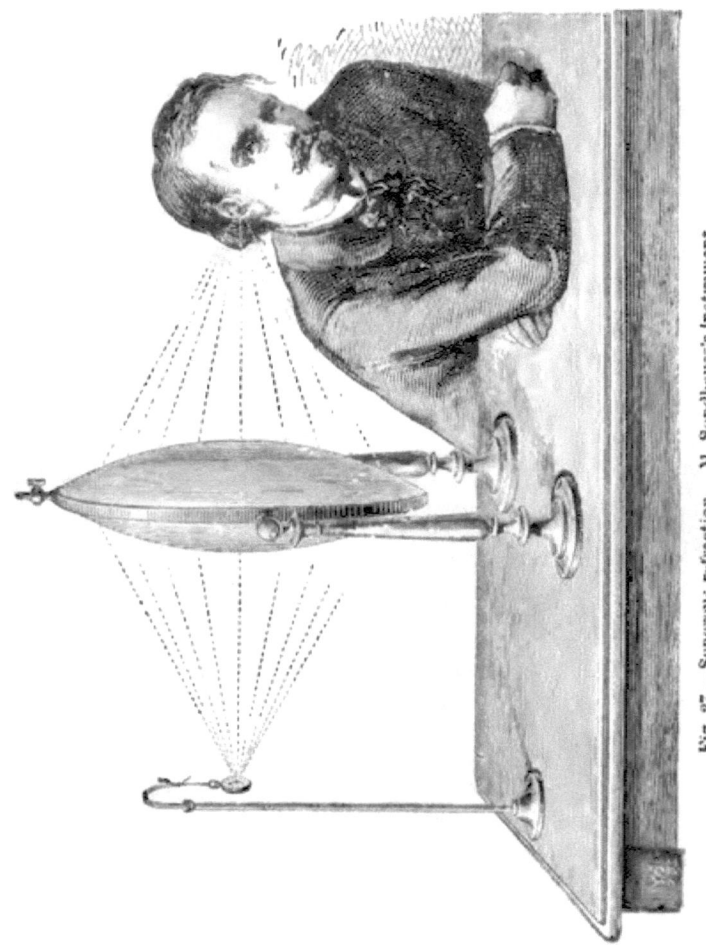

Fig. 27.—Sonorous refraction. M. Sondhauss's Instrument.

metres. If the distance of the reflecting surface exceeds half this distance, we are able to hear separately the returning vibration. If there be parallel reflecting surfaces at a distance, the echo becomes multiple instead of single. Buildings, rocks, clumps of trees, even clouds can produce such reflection, as in the case of thunder.

The following quotation fairly sums up the best known of these phenomena.[1] "In ancient and modern works a number of multiple echoes are mentioned, the surprising effects of which may be questioned, but which are all easily explained by successive reflection.

"Such an echo is said to have existed at the tomb of Metella, the wife of Crispus, which repeated a whole verse of the *Æneid* as many as eight times. Addison speaks of an echo which repeated the noise of a pistol-shot fifty-six times, like that of Simonetta in Italy. The echo of Verdun, formed by two large towers about fifty-two metres apart, repeats the same word twelve or thirteen times. The great Pyramid of Egypt contains subterranean chambers connected by long passages, in which words are repeated ten times. Barthius speaks of an echo situated near Coblenz, on the borders of the Rhine, which repeats the same syllable seventeen times, with a very peculiar effect, the person speaking being scarcely heard, while the repetitions produced by the echo are very distinct sounds. Among echoes in England may be noted one in Woodstock Park, which repeats seventeen syllables by day, and twenty by night; while in the Whispering Gallery of St. Paul's, the slightest sound is answered from one side of the dome to the other."

Reflection from Gases.—Reflection may also take place from layers of gases possessing different densities, a fact which has been studied by Tyndall. Sound from a high-pitched reed being conducted through a tube towards a sensitive flame serving as an indicator, was cut off by the interposition of a coal-gas burner of the ordinary "bat's-wing" kind; and by holding the latter at a suitable angle, the sound could be reflected from the flame, through another tube, in sufficient quantity to excite a second sensitive burner.

On account of the great difference of density reflection is nearly total at the boundary between air and solid or liquid matter. Hence sound in air is not easily communicated to water, and sounds whose origin is under water are heard with difficulty in air.

Sound Shadows.—When waves of sound impinge upon an obstacle, a portion of the motion being thrown back as an echo, there is formed under cover of the obstacle a sort

---

[1] Guillemin's *Forces of Nature*, p. 140.

of sound-shadow. To produce this in anything like optical perfection, the dimensions of the intervening body must be considerable. The standard is the wave-length of the vibration, and it requires almost as extreme conditions to produce *rays* in the case of sound, as in optics to avoid producing them. Still, sound-shadows thrown by hills or buildings, are often tolerably complete.[1]

Refraction of Sound by the Atmosphere is produced (1) by temperature, or (2) by wind.

1. The deviation of sonorous rays from a rectilinear course, due to heterogeneity of the atmosphere, has practical interest. The change of pressure at different levels does not give rise to refraction, since the velocity of sound is independent of density; but, as Reynolds has pointed out, the case is different with variations of temperature as usually met with. These are determined chiefly by the rarefaction or condensation which a portion of air must undergo in its passage from one level to another. Thus acoustical refraction dependent on temperature has almost the same explanation as that of the optical phenomenon of mirage. In the normal state of the air a ray starting horizontally, turns gradually upwards, and at a sufficient distance passes over the head of an observer on the same level as the source. The sound is heard, if at all, by diffraction. The observer may be said to be situated in a sound-shadow, though no obstacle may intervene.

The refraction is increased when the sun shines, and diminishes during rainfall.

2. It has long been known that sounds are generally better heard to leeward than to windward of the source, but Professor Stokes first showed that increasing velocity of wind must interfere with the rectilinear propagation of sound-rays. When the wind increases overhead, a horizontal ray travelling to windward is gradually bent upwards; rays travelling with the wind, on the other hand, are bent downwards, so that an observer to leeward hears by means of a ray which starts with a slight upward inclination, and which has the advantage of being out of the way of obstructions for the greater part of the course.[2]

The results of Reynolds's experiments were—

1. When there is no wind, sound proceeding over a rough surface is more intense above than below.

2. If the wind-velocity be greater above, sound is lifted to windward, and not destroyed.

[1] Rayleigh, II. p. 107.   [2] Ibid, *loc. cit.*

3. Under the same conditions it is brought down to leeward, and its range is extended at the surface of the ground.

Atmospheric refraction has been much studied with reference to fog signals at sea. Tyndall has moreover shown that sound may be intercepted by alternate layers of gases of different density. It is probable that both causes are concerned in the capricious behaviour of these warnings. Lord Rayleigh, moreover, points out that there is a difference in behaviour between long and short sounds. This agrees with Tyndall's observation that in some states of the weather a howitzer firing a 3 lb. charge commanded a larger range than the whistles, trumpets, or siren, while on other days the inferiority of the gun to the siren was demonstrated in the clearest manner.

Influence of Fog.—It has generally been believed, on the authority of Derham, that the influence of fog was prejudicial to the dispersion of sound. Tyndall proved that this opinion is erroneous, and that its passage is favoured by the homogeneous condition of the atmosphere which accompanies fog. When the air is saturated with moisture, the fall of temperature with elevation is much less rapid than in dry air, on account of the condensation of vapour which accompanies expansion. From a calculation of Thomson's[1] it appears that in warm fog the effect of evaporation and condensation would be to diminish the fall of temperature by one-half. The acoustical refraction due to temperature would thus be lessened, and in other respects the condition of the air would be favourable to the propagation of sound, provided no obstacle were offered by the suspended particles themselves.

[1] Manchester Memoirs, 1861-2.

CHAPTER III.

INTENSITY, CONSONANCE, INTERFERENCE.

Intensity of Sound.—The waves of rarefaction and condensation issuing from a sonorous body in a homogeneous medium, like the "rays" of light proceeding from a candle, must not be regarded as moving merely in a linear direction. It is true that both in the case of sound, and in that of light, the communication between the producer and the recipient takes a linear form; but the real constitution of the unconfined sound-wave is spherical. There being nothing to impede the oscillation of the ultimate particles, each impulse spreads in an enlarging and concentric shell, the quantity of matter set in motion augmenting as the square of the distance from the source. The *intensity*, or loudness, must therefore diminish in the same ratio. This is termed the law of Inverse Squares, and is true also for light. The small space through which each particle vibrates backward and forward is termed the *amplitude* of its undulation, and the intensity of sound is proportional to the square of this amplitude.

If the sonorous wave be confined in a tube, of course its progressive extinction by transference of motion to rapidly increasing masses of matter does not take place, and it may be conveyed for long distances with only very slight enfeeblement. On this principle are constructed the ordinary speaking-tubes. M. Biot, in the experiments by which he determined the velocity of sound in solid bodies, proved the fact that sound transmitted by the air in the waterpipes of Paris was not sensibly enfeebled at the distance of nearly a kilometre. Two persons speaking in whispers could easily hold a conversation through these pipes. "There is

III.] INTENSITY, CONSONANCE, INTERFERENCE. 63

only one way not to be heard," says M. Biot; "not to speak at all."[1]

There is, however, an important difference between the propagation of sound in a uniform tube and in an open

Fig 28.—Propagation of a sonorous wave through an unlimited medium.

space. In the former case, the layers of air corresponding to successive wave-lengths are of equal mass, and their move-

[1] Quoted in Guillemin's *Forces of Nature*. Regnault found the report of a pistol in a pipe of 1·10m. to be audible at a distance equivalent to 10,000 metres.

ments are precisely alike, except in so far as they are interfered with by friction. Regnault found that in a conduit of ·108 of a metre in diameter, the report of a pistol charged with a gramme of powder ceased to be heard at the distance of 1,150 metres. In a conduit of ·3m. the distance was 3,810m. In the great St. Michel sewer of 1·10m. the sound was made, by successive reflections, to traverse a distance of 10,000 metres without becoming inaudible. In an open space, each successive layer has to impart its own energy to a larger layer; hence there is continual diminution of amplitude in the vibrations as the distance from the source increases. An undulation involves the onward transference of energy; and the amount of energy which traverses, in unit time, any closed surface described about the source, must be equal to that which the source emits in unit time. The intensity therefore follows the same law as that of radiant heat, and of light, as stated above. The energy of a particle executing simple vibrations in obedience to elasticity, has been said to vary as the square of the amplitude of its vibrations; for the amplitude being redoubled, the distance worked through, and the mean working force are both doubled, so that the work done is quadrupled. At the extreme positions all is potential energy; in the middle all is kinetic energy; at intermediate points it is partly in one form and partly in the other. If we sum up the potential and also the kinetic energies of all the particles constituting a wave, we shall find the results to be equal.[1]

This assumption is not absolutely true; since vibration implies friction, and friction implies the generation of heat. Sonorous energy therefore diminishes more rapidly than according to the law of inverse squares, and, in becoming extinct, is converted into heat.

Mayer has devised a plan by which the intensities of two sounds of the same pitch may be directly compared. The two sounds are separated by an impervious diaphragm, and in front of each is a resonator accurately tuned to them. Each resonator is attached by caoutchouc tubes of equal length to a U-tube, in the middle of which is a branch leading to a manometric capsule.

If the resonators are at the same distance from the sounding bodies, and one be excited, the attached flame vibrates. If both are produced in the same phase and intensity they interfere completely in the tube, and the flame is stationary.

[1] Everett's Deschanel, p. 799.

III.] INTENSITY, CONSONANCE, INTERFERENCE.

If they be **not of** the same intensity, the interference will be incomplete, and the flame will vibrate. If one be then altered until the flame is again still, the intensities will be directly as the squares of their distances from the resonators. This instrument is therefore the correlative of Rumford's shadow Photometer.

Tabular Statement of Intensity.

1. Intensity inversely as square of distance.
2. Intensity directly as square of amplitude of **vibrations**.
3. Increases with density of medium.
4. Modified by motion of atmosphere.
5. Strengthened by proximity of sonorous body.

Intensity, force, or loudness, may be looked upon as the first characteristic of musical tone : *Pitch*, dependent solely on the rapidity of the vibration, is the second, and will be considered in the next chapter. *Quality* or character has been shown to be connected with the form of the vibration, and will be adverted to farther on.

Consonance.—A remarkable property of vibratory motions is the power they possess of communicating themselves to matter in their immediate neighbourhood. Even in a mechanical view of the subject this property is evident. If two pendulums, attached to different clocks, be fastened to one board and set going, it is well known to clockmakers that one will coerce the other into a spurious synchronism, which ceases directly they are divided. A regiment of soldiers crossing a suspension bridge, if keeping step and marching order, communicates regular impulses to the fabric of the bridge, and may even cause such oscillation as to endanger the structure; the swinging of the bells in a tall tower, such as that of Magdalen College at Oxford, itself produced by a succession of small impulses conveyed to the larger mass of each bell, is farther transmitted to the elastic material of the tower, producing in it very distinct oscillatory movements.[1] This property is even more noticeable in the swifter alternations which form a musical note. Whatever be

---

[1] "Illustrations of the powerful effects of isochronism," says Lord Rayleigh (*Theory of Sound*, p. 61), "must be within the experience of every one. They are often of importance in very different fields from any with which acoustics are concerned. For example, few things are more dangerous to a ship than to lie in the trough of the sea, under the influence of waves whose period is nearly that of her own rolling."

the source of sound, its effect is immediately transmitted to the particles in contact, and with an amount of force which at first seems disproportionate to its inherent energy. For although the third law of Newton respecting the equality of action and reaction must obviously be fulfilled, the elasticity of most bodies enables them to take up transmitted vibration in a very high degree. Those which possess this property in the most marked manner are called sonorous, and their responsive vibration is termed consonance. Without consonance the effect of musical sound would be slightly, if at all, appreciable, for it is by this means that its chief propagation and dispersion is effected. In the first rank as consonators stand the producers themselves. A tuning-fork is set into sympathetic vibration by another vibrating in unison with it. A string will perform the same office, and an organ pipe instantly reinforces the sound of a corresponding tuning-fork held near its open extremity. Even a jar or bottle, the cavity of which bears some definite ratio to the wave-lengths of the sounding body, answers a similar purpose.

The weight and density of the consonant body do not necessarily prevent its acting as a propagator of sound if its modulus of elasticity be high. Lead or clay for instance deaden sound by their inertness, while steel and glass convey

Fig. 29.—M. Helmholtz's resonance globe.

it with the utmost facility. But bodies of lighter character and less dense molecular construction, such as the softer woods, are obviously the fittest for this function. It is to the highly resonant structure of pine-wood that the predominant tone of the violin family especially is due.

Generally speaking, reinforcement in **sound** is correlative with the power of producing **it.** All sounding bodies reinforce, but some have been divided **off into** what Clerk-Maxwell **terms distributors.** Others have the power of singling out particular sounds for reinforcement. If, for example, the dampers be lifted off a piano and the voice be used in its neighbourhood, it will be heard to sing out loudly with a humming tone the notes which have been spoken on. The same effect occurs with drums and tuning-forks: even the flat crown of a hat responds by vibration sensible to the touch when **loud noises** occur in proximity to it.

This power **of** singling out sounds has been utilized by Helmholtz for the analysis of musical notes **in** making *Resonators.*[1] They originally had external membranes, but he afterwards found that the tympanic membrane, or drum of the ear itself, could be used for the same purpose, by making the resonant cavity of a particular size, such that, itself **speaking** a certain note, it will single out that note from all others, and reinforce it vigorously.

The simplest method, however, of demonstrating resonance **is** to take a tall jar or **tube** and hold over it a sounding body, such as a tuning-fork. As long as the fork and the cavity of the jar are in no definite relation to one another, the sound is unaltered. But if, by gradually pouring in water, we alter the depth of the cavity, a point is suddenly reached at which the note starts out with exceeding clearness. It will then be found that the length of the column of air in the tube bears **an exact** proportion to the wave-length of the vibrations emitted by the fork, usually that of one to four. The reason of this is obvious. At each vibration of the fork, a wave of condensation travels down the tube, is reflected from the bottom, and returns to find it in the same phase as when it **started.** It thus superposes its own motion upon that of the fork, and by a succession of such actions reinforces the sound. In so doing, however, the fork has the additional labour imposed **upon it of** setting in motion the contained particles of air as **well as its** own, and therefore comes sooner to rest than when vibrating independently. An effective experiment is produced by combining a sonorous bell with a resonant cavity of variable dimensions. A **source** of sound may also act upon a tuning-fork by consonance. If two **forks** in

---

[1] See Chapter V.

accurate unison be placed at some distance from one another, and one be excited, the other immediately begins to sound with vigour, and if the first be damped, it may be again set in motion by the continuance of derived vibration established in the second. Tuning-forks and other sonorous bodies, such as glass or metal vases, often commence sounding spontaneously when a musical instrument, an organ or harmonium, is played on in the same room ; even the glass windows of a church are apt to take up the note of a particular pedal pipe to the exclusion of those in its immediate neighbourhood.

Helmholtz has shown that a stretched string may be made to perform the office of a resonator.[1] "If a sounding tuning-fork have its stem placed on a string, and it be moved so near the bridge that one of the proper tones of the section of string lying between the fork and the bridge is the same as that of the tuning-fork, the string begins to vibrate strongly, and conducts the tone of the tuning-fork with great power to the sounding-board and surrounding air ; whereas the tone is scarcely if at all heard as long as the section is not in unison with the tone of the fork."

A simple apparatus was used by Savart to show the influence of jars or boxes in strengthening sound. Close to a source of sound, such as a bell or tuning-fork, was placed a hollow cylinder, closed at its farther end by a moveable bottom, by means of which its capacity could be increased or diminished. This was supported on a sliding rest, so that its open end could be brought near or removed from the vibrating body. The bell or fork being excited by a rosined bow, the cavity of the resonator was altered until it coincided in pitch with it, and immediately the sound, originally feeble, and all but inaudible, became distinct and loud. The loudness could then be varied to any extent by moving the open end of the consonating cavity into closer proximity to the source of sound. Helmholtz has utilized the latter phenomenon in his synthetical reproduction of compound vowel-qualities. Kœnig has improved on the original form of resonator by introducing a slide such as that named above, by which the same instrument may be made to reinforce several notes.

Theory of Resonators.— In a pipe closed at one end, we have a mass of air vibrating in certain definite periods peculiar to itself, in more or less complete independence of the external atmosphere. If the air beyond the open end were

---

[1] Helmholtz, *Sensations of Tone*, Ellis's translation, p. 88.

entirely without mass, the motion within the pipe would have no tendency to escape, but in actual experiment the inertia of the external air cannot be got rid of. When the diameter of the pipe is small, the effect of this is small, and vibrations once excited have a certain degree of persistence. The narrower the channel of communication between the interior cavity and outside, the greater does the independence become. In the presence of an external source of sound, the contained air vibrates in unison, with an amplitude dependent on the relative magnitudes of the natural and constrained periods, rising to great intensity in the case of approximate isochronism. When the original cause of sound ceases, the resonator yields back the vibrations stored up within it, thus becoming itself for a short time a secondary source. A vessel containing air, of which the capacity is sufficiently large, communicating with the external atmosphere by a narrow neck represents a state of things in which the kinetic energy may be neglected except in the neighbourhood of the aperture; the air moving approximately as an incompressible fluid would do under like circumstances, sufficiently so for a calculation of the pitch.[1]

Sound-boards generally as Consonators.—It is obvious that the vibrations of strings, tuning-forks, reeds, and other generators of sound, cannot, unaided, impress any large amount of motion on the surrounding air; consequently the tones they produce alone are very feeble. It has been shown how these may be increased for individual notes. There is however a large class of appliances with which most musical instruments are furnished, termed sound-boards. These do not so much tend to intensify particular rates of vibration, as to impart sonority to all. Their function has been well explained by Stokes; whatever their absolute size, they present a far larger area to the air than the string or fork itself. By thus laying hold, as it were, of the rarer body, they prevent the dissipation of the alternate rarefactions and condensations of which a sound-wave consists. Sound-boards usually take the form of large surfaces or boxes of some elastic material, such as pine-wood. The older pianos had a flat sheet of varnished deal laid immediately above, but not touching, the strings. The most remarkable example, however, occurs in the body of all instruments of the viol or violin family. We have here a flattened box of a very complicated shape, the front or belly of which is made of fine even-grained pine-wood; the back

[1] Rayleigh, *On Sound*, II. 156.

usually of a harder and more compact material. These two are united by a bar of similar wood termed a sound-post, and the former is pierced by two openings, usually in the shape of the letter f. Upon the belly stand the feet of the bridge, and close to one of these feet is the sound-post. No better combination could have been adopted for securing the largest possible amount of resonance. The wood itself has been shown to be the most rapid and perfect of all sound-transmitters; its surface is large proportionally to the size of the instrument; the two faces are rigidly connected by a conductor, and a considerable mass of air is inclosed, communicating by the two orifices above named with the surrounding air. This is perhaps the most perfect instance of the fact named in the introductory remarks, of a machine which has grown up fortuitously, the result entirely of artistic experience, but which proves to conform in the most perfect way to theories subsequently developed.

Interference of Sound may be looked upon as the logical correlative and contrary of consonance. It has already been shown how two waves of similar period and phase reinforce one another; and it naturally follows that if they be of unequal period or opposite phase they must be mutually destructive. If the phase alone differ, the period of the two being equal, the interference is constant; if, however, the periods are slightly unequal, alternate reinforcement and interference takes place, giving rise to the phenomenon of Beats. As in the case of consonance, this action can be followed up from the mechanical department in which pendulums, and even waves in water are types, through sound, into the far subtler and more rapid undulations which characterize light. If two waves pass simultaneously through the same medium, the actual motion of each particle is the result of the two combined. This will be the sum of them if they be in the same direction, the difference, if in opposite directions, and nothing if they be equal and opposite. This statement is true whether the undulations be on the surface of water, in air, or in the æther which is believed to transmit light. Reducing it to the terminology of sound, two undulations of equal wave-length and amplitude must either be in the same phase or always differ by the same amount; if the same, the result is a doubling of the sound, and is a case of consonance: if opposite, that is if one be exactly half a wave-length in advance of the other, the two motions will be opposite, and silence will result from the conjoint action of two sounds. This is the case of complete interference; a

INTENSITY, CONSONANCE, INTERFERENCE.

condition which can, however, be only partially realized in practice. The interferences can also be produced between a direct and a reflected wave, and were carefully experimented on by Savart and Seebeck, using a distant wall for the reflector.

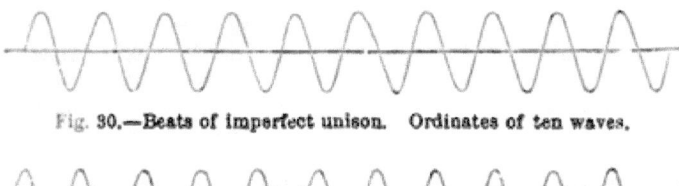

Fig. 30.—Beats of imperfect unison. Ordinates of ten waves.

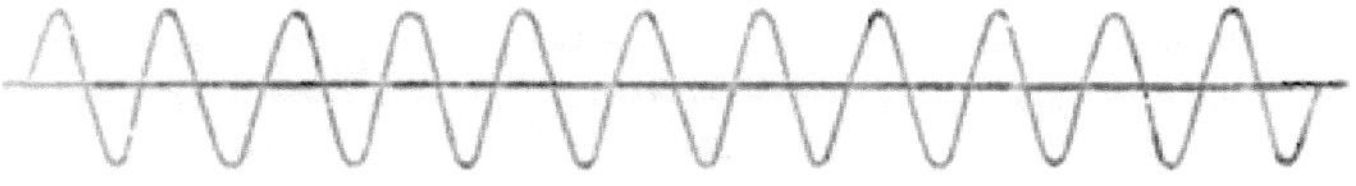

Fig. 31.—Ordinates of eleven waves transmitted in the same time.

The simplest illustration **of interference is** afforded by an ordinary tuning-fork. If **it be set** in vibration and held to the ear, its note will **as a rule be** easily heard; but if it be slowly rotated on its stem, it will be found that there are four positions in each rotation in which the note is distinctly audible, and four others in which it disappears altogether. It is heard whenever its two prongs are held parallel to the ear, and also when they stand at right angles to that plane. Between these are **four** oblique situations almost at an angle **of 45°** with those previously named, which are positions of **silence**. The sound of a tuning-fork is **at** all times much diminished **by** the opposite motion of the **two** prongs, which tends **to** cancel the vibration of the surrounding air. This may be shown by passing **a** small tube over one prong, without touching it, and thus shielding the other from its antagonistic influence; the sound immediately becomes materially stronger. When **on** the other hand, the fork is held edgewise to the ear, the prong next it, being in the more favourable position of the two, exercises a predominant influence, and enables the sound to be heard. But when the oblique position is obtained, **a more** complete antagonism **results** and the note is entirely extinguished. This experiment **may** be made more distinct by rotating the fork over the **mouth** of **a** consonating jar or tube.

If two perfectly similar stopped organ-pipes be **set on the** same wind-chest, and either be singly excited, **a** note **is** produced **in the** ordinary way. But if **the two are** blown together, interference takes place between them, and nothing is heard beyond the rushing of the wind.

If a circular disc be excited by means of a bow, so as to produce six vibrating sectors, and the palm of the hand be brought over one of these, the sound is at once intensified. This does not occur if the two hands are placed over two consecutive portions, but is more marked when alternate sectors are so treated. The hands, acting as dampers, check the spread of the vibrations from those parts of the plate immediately beneath them, and allow the opposite motion of the intervening plate to develop, comparatively free from interference, into sound. The experiment is even more striking if, instead of the hand, a piece of pasteboard with three alternate sectors cut out of it be employed as the damper. It has been also varied by fixing a tube bifurcated at its lower extremity, and furnished at its upper with a vibrating membrane over the plate. Sand is scattered on the membrane; when the two orifices of the bifurcation are held over adjacent sectors, little or no motion of the membrane is seen, but when over alternate sectors, the sand is immediately thrown off; in the first case there is interference, in the second coincidence of vibration.

Perhaps the most beautiful illustration is one proposed originally by **Sir John Herschel**. A brass tube divides into two branches, which reunite further on in their course. One branch can be drawn out by means of a slide to a greater length, so that the waves from a sounding body can be made to travel over different lengths of tube. If a vibrating tuning-fork be brought near to one orifice, and the ear applied to the other, as long as the two tubes are of the same length the note is clearly heard. But on drawing out the slide, a point is reached at which the sound vanishes. This takes place when the extended branch is half a wave-length longer than the other; indeed it affords a rough method of measuring wave-lengths.

Beats.—The most usual form in which interference is met with is that of beats. These generally originate in the simultaneous sounding of two notes not quite in unison. They will be shown later to be our most subtle and trustworthy method of securing and testing unison or perfect concords. They are easily explained on the principle of interference. The wave-lengths of two notes being slightly different while the velocity of propagation is the same, the phase will alternately agree and disagree in their course. When the phases of the two waves are coincident they strengthen one another, when the phases **are** opposite they destroy one another. If a C tuning-fork of 528 vibrations

Fig. 32.—**Sums** of the corresponding ordinates.

in a second be combined with one slightly sharper **giving** 529, an exact second will intervene between every **successive** interference and reinforcement; for in this time the slower wave has been gradually falling behind the quicker, until the loss amounts to a whole wave-length. The number, therefore, **of** beats is the difference in the frequencies of vibration of **the** beating notes. It will be observed that beats as a mode **of** measurement are not dissimilar in principle to the Vernier which is employed in astronomical instruments.

Beats may be demonstrated **by means** of **two** similar tuning-forks, a prong of one having been loaded **with a** little wax so as to alter slightly its period of vibration; **or by two** organ-pipes, one of which is slightly "shaded" by **the hand** either at the mouth or at the open end. Two strings **on the** same sonometer to which slightly different tensions **are** applied produce loud beats. But they can be shown at **once** to the eye and the ear by means of two singing-flames. **To** the tube surrounding one of these a slider is adapted **by** which its **length** may be varied; if this be moved until unison is attained **the** two flames sing steadily and without flickering; or continuing the movement of the slider, the beats recommence and **the** flames dance up and down in time to them. For reasons stated later it will be seen that free reeds such as those of harmoniums give singularly distinct beats. For this cause, among others, they are specially valuable as standards of pitch.

Accidental illustrations of **beats** may be heard whenever a large organ tuned according **to** equal temperament is **played.** They are very audible in the sound of large bells, from unavoidable want of symmetry in the figure. A good **mechanical** example of interference may be noted when the big **bell at the** House of Commons tolls the hour. The first blow of the hammer falls **on the** metal in a quiescent state, **and a** sound of medium loudness is elicited. But as the clockwork lifts and drops its weight at regular periods, before the first vibration is extinguished, two conditions may occur; either the hammer may meet the bell in the same phase as its own, in which case an extremely loud toll results, or it may fall on it in the opposite phase, when great part of the momentum is employed in cancelling the interfering vibrations already set up and a very feeble sound is given out. These differences of intensity in the successive strokes can be plainly heard at 11 o'clock or midnight when the air is still.

CHAPTER IV.

PITCH.—ITS MEASUREMENT, LIMITS, VARIATION, STANDARDS, AND TONOMETRY.

It has been already stated that all sound consists of three elements, namely, **intensity, pitch,** and **quality.** The first of these depends entirely on the amplitude of the vibrations; the last has been shown by Helmholtz to be connected with certain secondary and affiliated oscillations termed harmonics, from which few musical tones are entirely free. This will be adverted to in a subsequent chapter. The remaining constituent, namely, pitch, has been the subject of much important research. It depends entirely on the number of vibrations in a given time.

Limits of Audible Sound.—Savart showed that the faculty of perceiving sounds depends rather on their intensity than on their acuteness, and by increasing the diameter of his toothed wheel, carried it up to 24,000 vibrations per second.

For deep sounds he substituted for the toothed wheel a bar about two feet long, revolving on an axis between two thin wooden plates about 0·08 of an inch distant from it. A grave, continuous, very deafening sound was thus produced, with 7 to 8 vibrations in a second. These results are disputed by Despretz and Helmholtz, the former placing the limit at 16, the latter at 30 vibrations, the definite musical character according to the latter observer being only obtained at 40 vibrations per second. These discordant results are no doubt due to the different capacities of different observers for the perception of sound, indeed the extreme upper limit of audible sound appears to vary materially with the individual. M. Despretz had a diatonic octave of small forks from 8,000 to 16,000 vibrations in the second, tuned by M. Marloye, who declared that with practice he could tune still higher scales.

He did tune an octave fork to that last named, which would give 32,000 vibrations. He states that in the process of tuning he went twelve times over the whole octave. On the two first attempts he heard nothing. At the third attempt he was able to distinguish the intervals in the following order: Fourth, Fifth, Minor Sixth, Minor Third, Major Sixth, Major Third, Minor Seventh, Major Seventh, and at last with great difficulty, the Major Second. Upon this he makes the suggestive remark that if there be a natural scale for the ear, these observations would point to its being minor rather than major.

Appunn has made thirty-one tuning-forks, in true major scales from 2,048 vibrations up to 40,960, which most ears can distinguish, although they often produce a very painful sensation.

Captain Douglas Galton has shown a method of producing even higher tones, by means of small whistles. Many of these notes appear to be more audible to the smaller mammalia, especially to cats, than they are to the human ear.

Preyer has made experiments of considerable precision by means of which he fixes the minimum limit for the average ear between 16 and 24 single vibrations per second, the maximum at 41,000; many persons of fair hearing powers being however deaf to sound of 16,000 or even fewer vibrations.

It appears, from experiments made by the writer, that the musical character of low tones depends materially upon the presence or absence of a sufficient consonant body for their reinforcement and co-ordination. Sixteen-foot C can be obtained on the double bass, with distinct musical character, by special treatment of the resonant body of the instrument. The size of the room, moreover, must be considerable for the large waves thus originated to spread without damping and interference. The 32-foot octave of pedals in the organ at the Albert Hall is perfectly musical in effect. Probably the experiments of Helmholtz, which were made with feeble sources of sound such as metal strings, were deficient in the requisites just named. The lowest and grandest note to be heard occurs when a train passes into a short tunnel. The successive explosions of issuing steam, varying in rapidity from 8 to 20 in a second, can, if the speed be gradually increasing, be clearly heard to coalesce into a profound humming note of steadily rising pitch due to the consonance of the gigantic resonator furnished by the bore of the tunnel itself.

PITCH.—ITS MEASUREMENT, ETC.

The determination of the number of vibrations in a given period corresponding to a particular musical note may be made by the various methods of Tonometry, as this branch of acoustics has been termed. These may be given, in tabular form, as follows:—

I. Mechanical methods.
1. Savart's toothed wheel.
2. Cagniard de Latour's siren.
3. Perronet Thompson's weighted monochord.
4. Duhamel's vibroscope.
5. Leon Scott's phonautograph.
6. Edison's phonograph.

II. Optical methods.
1. Lissajous' method.
2. Helmholtz's vibration microscope.
3. Kœnig's manometric flames.
4. McLeod and Clarke's cycloscope.

III. Photographic methods.
Prof. Blake's experiments.

IV. Electrical methods.
1. Meyer's electrical tonometer.
2. Lord Rayleigh's pendulum experiment.

V. Computative methods.
1. Chladni's rod tonometer.
2. Scheibler's *Tonmesser* with tuning-forks.
3. Appunn's tonometer with free reeds.

I. **Mechanical Methods.**—The simplest mechanical method in the above list of contrivances is founded on the fact that slight successive noises caused by collision, which individually present no musical character whatever, gradually coalesce into a definite tone if their intervals be regular and if their succession be sufficiently rapid. A common watchman's rattle, and even a stick passed rapidly across the bars of a grating, may be used as a popular illustration of the fact; Dr. Haughton has furnished an excellent example, named in the introductory remarks.

"The granite pavement of London is four inches in width, and cabs driving over this at eight miles an hour, cause a succession of noises at the rate of thirty-five in the second, which corresponds to a well-known musical note, that has been recognised by many competent observers; and yet nothing can be imagined more purely a *noise*, or less musical,

Fig. 33.—Savart's toothed wheel.

than the jolt of the rim of a cab-wheel against a projecting stone; yet if a regularly repeated succession of jolts take place, the result is a soft, deep, musical sound, that will well bear comparison with notes derived from more sentimental sources."[1]

(1) **Toothed Wheels.**—Even as early as the time of Galileo, that philosopher produced a musical sound by the passage of a knife over the serrated edge of a piastre and inferred from this that the pitch depended on the rapidity of the impulses. "On July 27, 1681, Mr. Hooke showed an experiment of making musical and other sounds by the help of teeth of brass wheels, which teeth were made of equal bigness for musical sounds, but of unequal for vocal sounds."[2]

Savart first reduced the system to accuracy by mounting the toothed or serrated wheel on an axis connected with machinery competent to give it rapid rotation, and attaching to it a counter or indicator showing the number of revolutions in a given period. Thus, giving the wheel 600 teeth, and rotating it forty times in a second, he could obtain 48,000 collisions in each second, which correspond to an extremely high note. A piece of card or a metallic plate was applied to the passing teeth, which of course received in a second as many blows as the product of the teeth into the rotations.

(2) **The Siren,** invented by Cagniard-Latour in 1819, substitutes for the successive collisions a series of small puffs of air. In its simplest form, as made by Seebeck, it consists of a rotating disc, perforated with circular rings of orifices, to each or any of which can be adapted a nozzle delivering wind at high-pressure from powerful bellows. In this form however the sound is feeble. The Siren, in spite of the quaint inaccuracy of its name,[3] was a considerable advance upon Savart's wheel. The teeth of the latter are here replaced by coincident openings in two similar circular plates, the one fixed, the other rotating above it, with but slight friction, upon an axis. In the act of rotation similar superposed rings of holes alternately open and close a passage for the wind issuing in a steady stream through the lower fixed plate. The isolated "puffs" soon unite into a continuous note.

In Cagniard-Latour's original instrument, a more complicated arrangement exists. The rotating disc turns on a vertical axis above a receptacle, the upper surface of which, in close approximation to the under surface of the disc, is pierced, not with a single hole, but with a ring of holes equal

---

[1] *Natural Philosophy Popularly Explained*, p. 157.
[2] Birch's *History of the Royal Society*. Quoted in Tyndall *On Sound*.
[3] It is said to have derived this name from its power of singing under water, a gift with which Homer's Σειρῆνες were not endowed.

80 ON SOUND. [CHAP.

in number and similar in position to those of the disc above them. Instead, however, of the holes in the two rings being pierced vertically both are inclined obliquely, the lower row

Fig. 34.—Seebeck's siren.

in one direction, the upper in an opposite sense. The object of this is to furnish motive power to the rotating disc from the horizontal element of the wind-pressure. It would have been better to omit this device, and rotate the disc by external force, as will be presently described. The upper end of the spindle carrying the disc is furnished with an endless screw, which works into a small registering train of wheels. These can be thrown in and out of gear at will; so that the rotations in any given number of seconds can be approximately indicated on dials outside. The indication is only approximate, since the added friction of the train, however small, tends to slacken the speed of the disc, and lower the pitch of the note when it is applied.

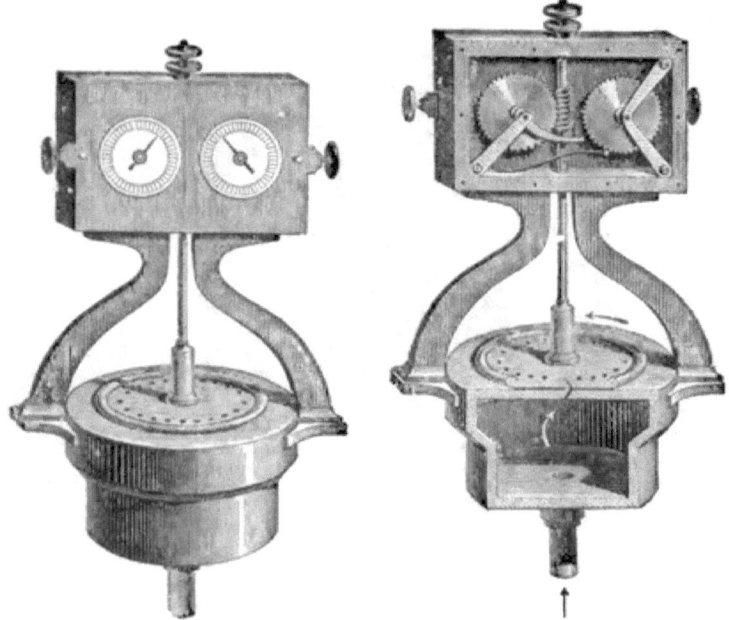

Fig. 35.—Cagniard de Latour's siren.   Fig. 36.—Interior view of the siren.

A very superior form of the siren has been introduced by Helmholtz, founded on the polyphonic siren of Dove, in which several notes can be sounded together. Helmholtz's instrument consists of two superposed sirens with their respective discs adapted to the same axis or spindle. The two wind-chests are fixed, one below the lower, and the other above the corresponding upper disc. The upper wind-chest with its attached orifices can be rotated on its axis by means of a toothed wheel and handle, whereas the lower receptacle is firmly fixed.

On each of the two discs are four rows of holes, which can be blown separately or in combination. The lower sets consist of 8, 10, 12, 18 holes, the upper of 9, 12, 15, 16. In order to damp the upper partial tones by means of a resonance chamber, cylindrical boxes in two sections are attached to the wind-chest by means of screws.

G

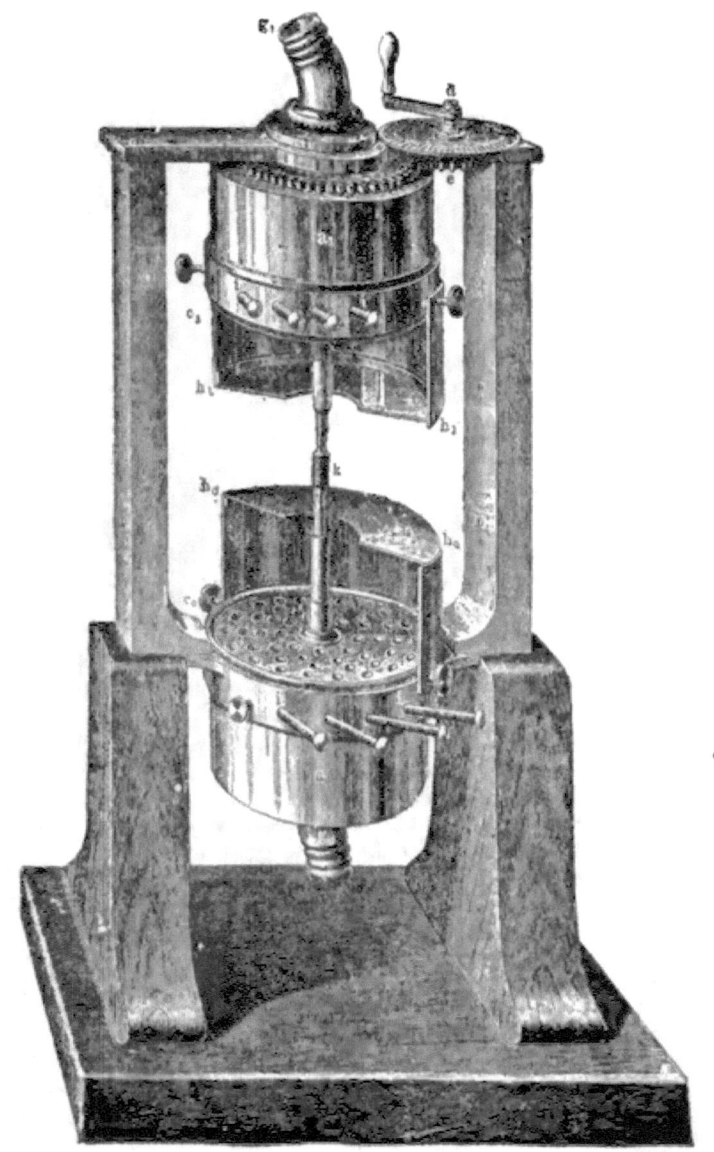

Fig. 37.—Helmholtz's double siren.

PITCH.—ITS MEASUREMENT, Etc.

The siren, although theoretically a perfect instrument, fails somewhat in practice, chiefly in consequence of the difficulty experienced in keeping its note steady. The character of the note itself is harsh and screaming, so that beats with softer sources of sound are all but inaudible. As it is usually made moreover, there is no way of preventing a steady acceleration of the rotation, or the corresponding rise in pitch. The blast of wind being made to accomplish two purposes, as a driving power as well as a source of sound, cannot be materially altered without at once reducing the impelling force and the tone. Mr. Ellis notes that "as each revolution of the disc reckons as twelve vibrations, an error of one revolution in a second, which is easily made, vitiates the results by twelve vibrations or ·4 of a semitone at the pitch of C, which is a large amount. Practically a siren cannot be depended on within ten vibrations."

Helmholtz, in whose hands the siren was made to give very fair results, employed an electro-magnetic driving machine to actuate it. It is connected with the discs by a thin driving-band. The siren does not then require to be blown. Instead of blowing, he places on the disc a small turbine constructed of stiff paper, which drives the air through the openings whenever they coincide with those in the chest. "This arrangement," he states, "gave me extremely constant tones on the siren, rivalling those on the best constructed organ-pipes."

Error of Siren.—Another source of error in the indications of the siren does not hitherto seem to have been noticed. This is due to the amount of compression to which the air is subjected. For properly driving the disc at high speeds very considerable force is necessary, on account of its inertia and friction. The wind in the chest should support a column of water from 12 to 24 inches in height, a pressure equivalent to from $\frac{1}{2}$ lb. to one pound per square inch. In passing through the perforations of the siren it is therefore altering materially in volume, and still more perceptibly in heat. Both these elements exercise a powerful influence on the tone emitted by wind instruments of all kinds, as will be shown in greater detail in a later chapter, and cannot be neglected in this instance with impunity.

Its Real Value.—The real practical use of the siren is for demonstrating the formation of the scale, and the vibration ratios which distinguish the principal concords and dissonances. These remain perfect and undisturbed in spite of variations in the absolute note upon which they are founded.

(3) Determination by the Monochord.—One of the earliest successful attempts at accurate determination of pitch was made by Perronet Thompson. For this end he revised and perfected the ancient instrument of Euclid and Pythagoras, the monochord. According to his construction it was five feet long, ten inches broad, and six deep; the wire was of steel the twentieth of an inch in diameter, containing 145 feet to the pound avoirdupois, breaking with a weight of 300 lbs. The load required to produce tenor C of the pianoforte was from 240 to 250 lbs. The sound was brought out by the application of a well-rosined bow, and had the strength of a violoncello. The method of using the above apparatus for the enharmonic tuning of an organ, will be described in a later chapter. Here it will be sufficient to note the direct physical method of measurement with such an instrument. A string is tuned to a given note, and its vibrations are determined by knowing the stretching weight, the weight of the wire as stretched, and the vibrating length of the string.[1] The following is the formula usually adopted, as given by Mr. Ellis in his excellent communication to the Society of Arts.

Let V = Pitch, or number of double vibrations in one second.

W = Number of grains in the stretching weight.

S = Number of grains in one inch length of stretched string.

L = Number of inches in vibrating string.

Hence SL = Weight of vibrating string; which, cut off, weighed and measured, gives L, SL, and S.

Then

$$V = \frac{\pi}{2} \sqrt{\frac{W}{SL} \times \frac{P}{L}}$$

P being the length of the seconds pendulum = 39·14 at Greenwich, and π the constant 3·14159.

The string is brought into sensible unison with the given note by shortening or lengthening the wire, and cut to the correct length. It is carefully measured for L, and weighed for SL. The weight with its attachments is weighed for W.

In this way Dr. Smith, in the year 1755, in the month of September, tuned a wire to give a note two octaves below the D pipe of the organ in Trinity College, Cambridge; arriving

---

[1] *Journal of Society of Arts*, May 25, 1877, "On the Measurement and Settlement of Musical Pitch," by Alexander J. Ellis, F.R.S., &c.

at the result of $D = 262$ or mean tone $C = 468\cdot7$ more than a whole tone below the usual pitch at the present time.

The above method, in spite of its theoretical beauty, is so liable to constructive difficulties that it is of little or no practical value for the determination of pitch.

A somewhat better form of the monochord for this purpose was introduced by Griesbach, and is preserved in the collection at South Kensington. It consists of a thick gut-string stretched over a body like that of a double bass. It was tuned two octaves below the note to be measured. Then a fine point being attached to one part of the string, a long strip of paper was passed over it at uniform velocity, and in passing was pricked by the point at every double vibration of the string. The notes being then counted and multiplied by four, the pitch of the fork was approximately determined. The employment of this method, also open to numerous sources of error, rendered the fork issued by the Society of Arts too sharp by ·37 of a semitone.

The monochord, although it produced good results in the hands of Perronet Thompson, for tuning correctly the different notes of the scale, is hardly so satisfactory as a means of determining absolute pitch. Scheibler sums up his long experience with it thus:—" Had it been possible to obtain exact results with a monochord, I could not but have succeeded, during the many years that I devoted to it, in tuning the forks of my scale correctly. My ear, and those of all others, were satisfied with the purity of the notes on instruments tuned by my monochord forks. But my mind would not be satisfied, because my results were not constant. When for example, one monochord showed me that a certain fork was one stroke of the pendulum too sharp, another monochord gave it as too flat. I became convinced that a mathematical monochord could not be constructed. I had also discovered that the string could not be protected from the warmth radiated by the observer's body, even when it was so thoroughly covered that there was only just space enough left for striking it. The string of a monochord, from this cause, does not remain for 30 seconds at the same pitch, but varies constantly by one-tenth to one-half of a double vibration."

In another place he estimates the possible error of the monochord at five double vibrations.

(4) **Graphic Methods** have the advantage of substituting a purely mechanical operation for a process requiring the assistance of an accurate musical ear. In their simplest form,

they may be typified by attaching a small point or style to the prong of a tuning-fork, and allowing this to trace its movements upon a piece of smoked paper or glass allowed to travel

Fig. 38.—Vibroscope.

steadily before it. If the fork be not sounding, the point will describe a straight line. But if it be first set in vibration, the attached point will constantly move backwards and forwards, and the wave-line remains as a permanent image of the motion performed by the fork during its musical oscillations. In practice it is best to wrap the paper round a rotating cylinder; rotatory movements being, as a rule, more easy to regulate, and steadier than those in a straight line. The cylinder may, moreover, be made to move in a spiral, instead of performing a simple axial rotation, and thus the line may be indefinitely extended within a limited space.

The curve obtained from any simple harmonic motion is one of those denominated curves of Sines; and it may, as Helmholtz remarks, be made to reproduce the motion of the vibrating points, by cutting a narrow vertical slit in a piece of paper, and placing it over the curve-tracing: if this latter be drawn uniformly under the slit from right to left, the point, seen through the slit, will appear to move backwards and forwards precisely in the same manner as the original tracer attached to the fork.

This method is susceptible of very considerable accuracy: and indeed has been employed for the exact determination of both its factors. Chronographs, in which a steadily vibrating tuning-fork is the standard, tracing its oscillations on a sheet of blackened paper carried past it, have been constructed: a fine example was exhibited at the Loan Exhibition of Scientific Apparatus. A second style, beside that on the fork, is usually made to note, by an abrupt motion on the same strip of paper, the exact instant of any astronomical or other phenomenon, such as the flight of a projectile, which it is desired to determine. On the other hand, the Phonautograph of Scott and the Vibroscope of Duhamel, which is practically the apparatus described above, aim at measuring the pendular vibrations of a sounding body in terms of the cylinder's rotation.

(5) **The Phonautograph.**—This instrument replaces the tuning-fork shown above, by a hollow barrel about 18 inches long and a foot in diameter. One end is open, the other is closed except by a tube carrying on it a stretched membrane. Upon the membrane is fixed a bristle, which moves with the membrane, and acts as a style. In order that this should not be at a node, the membrane is touched by a moveable piece, which, being made to touch the membrane first at one point and then at another, enables the observer to alter the arrangement of the nodal points. The bristle is thus made to coincide with a loop, that is a point at which the vibrations of the membrane are at a maximum.

88 ON SOUND. [CHAP.

When a sound is produced, the air in the cavity of the barrel and the membrane vibrate in unison with it, and the style is made to trace on a rotating blackened surface the

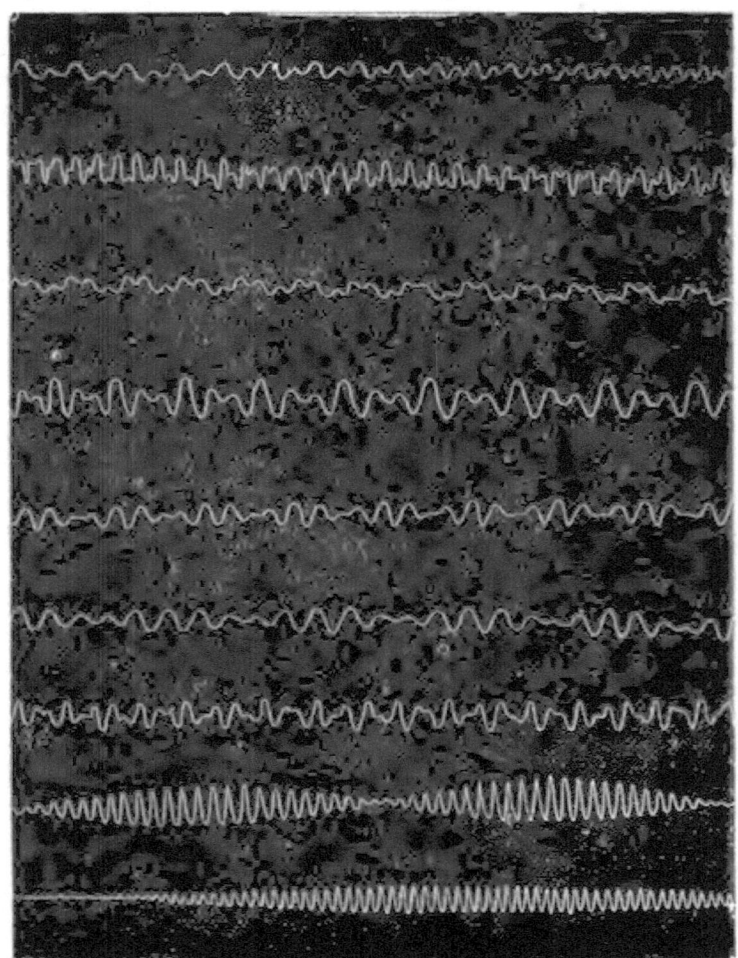

Fig. 39.—Combination of two parallel vibratory movements.

form of these vibrations. Each undulation corresponds to a double vibration of the style. The fork used in the vibroscope is retained, and traces beside the tracing of the style

IV.]   PITCH.—ITS MEASUREMENT, ETC.   89

another of regular shape and period, which forms an exact means of measuring the short intervals represented by the motion of the style.

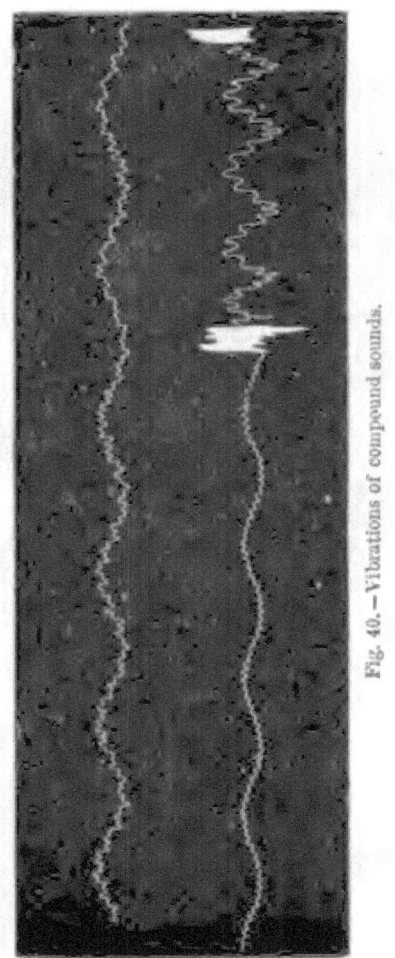

Fig. 40.—Vibrations of compound sounds.

(6) **Edison's Phonograph** retains the membrane and style of the above apparatus, as well as the spirally rotating cylinder of the vibroscope, but traces the impression on a

sheet of tinfoil, as will be more fully described in a later chapter.

II. (1) **Optical Methods.**—A very material improvement on the graphic methods just named has been made by substituting

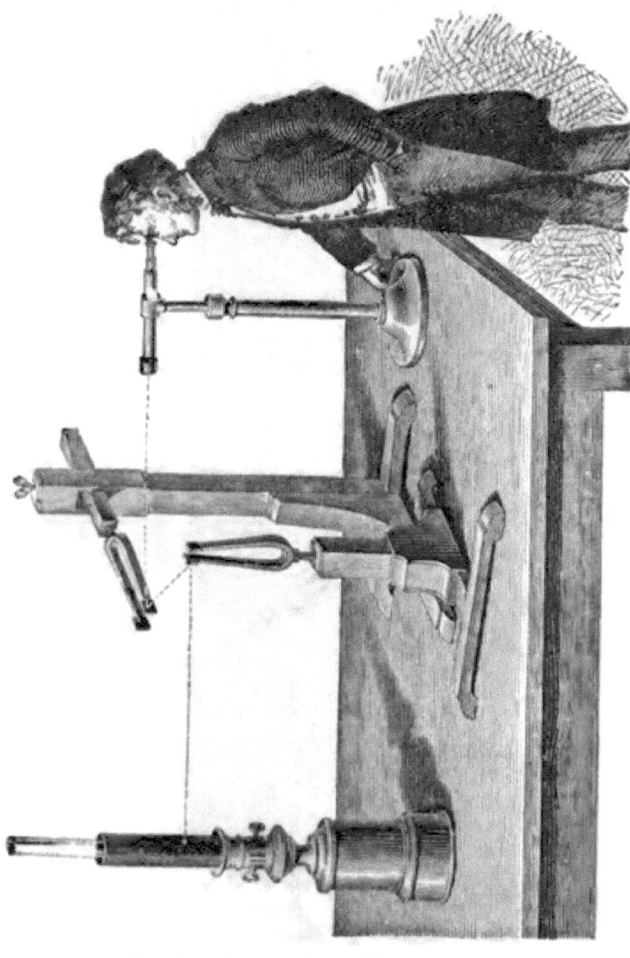

Fig. 41.—Optical study of vibratory movements.

a ray of light for the style. This method was originally due to M. Lissajous, who introduced a beam of light into a dark chamber, and, focussing it by means of a lens upon a mirror attached to one of the prongs of a vibrating tuning-fork, threw the image thus obtained upon a screen. The spot of light originally produced by the quiescent mirror is elongated on the fork being set into vibration, and forms a line of continuous illumination.

If either the tuning-fork carrying the first mirror, or a second mirror from which the beam is subsequently reflected, be moved through a small horizontal angle, the image will also move, and the line will be expanded into a sinuous figure in every respect reproducing the curve of Sines above described. The advantage of this method obviously is that the weight and friction of the style are entirely dispensed with. A second tuning-fork vibrating at right angles to the first can be made to carry the second mirror, and thus the composition of the two harmonic motions may be caused to produce the beautiful figures usually named after their original producer.

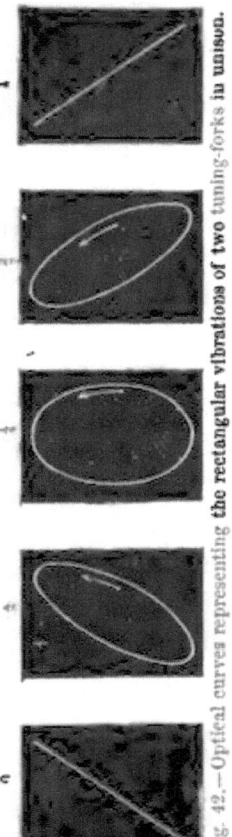

Fig. 42.—Optical curves representing the rectangular vibrations of two tuning-forks in unison.

An ingenious instrument of the same nature, but depending on the vibration of reeds instead of tuning-forks, was exhibited at the Loan Exhibition by M. Pichler. It consists of a wind-chest with means of blowing, above which are two reeds; one fixed in a vertical, the other in a horizontal direction; by shifting the bearing of one reed it may be made to increase its length, so that the vibrations of the two shall be to each other in any ratio from that of unison to that of an octave. They can pass through all intermediate figures. On each reed is placed a small mirror, and a beam from a strong source of light falls first on the mirror of the upper reed, whence it returns and is reflected on the second, and thence it is thrown on the screen. While the mirrors are motionless, the spot of light remains still; when they are set in vibration the one in a vertical position gives a vertical line of light; when the other in a horizontal plane is added they combine two harmonic motions, giving the curves already named. The circle denotes unison; the varied figures are

92   ON SOUND.   [CHAP.

produced by varying phases and velocities of the two reeds. The ear hears the different intervals at the same instant. Concord is thus denoted simultaneously by the absence of

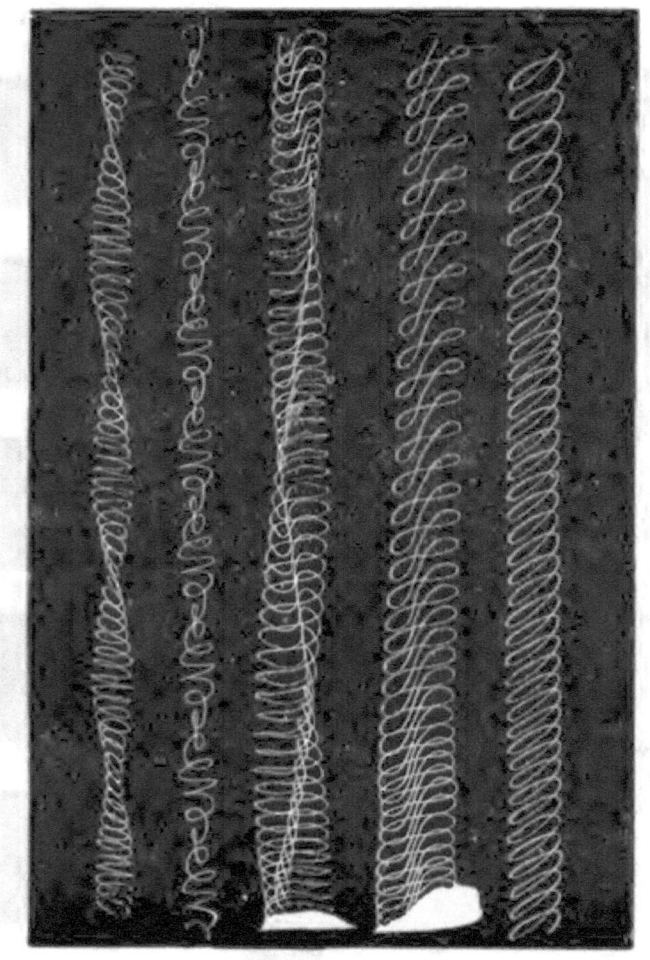

Fig. 43.—Combination of two rectangular vibratory movements.

beats, and by the stillness as well as the regularity of the resultant optical figure. Discord can be perceived by the rough clashing of the interfering undulations, and by the

IV.]   PITCH.—ITS MEASUREMENT, ETC.   93

flickering unsteadiness of the pattern thrown upon the screen.

The combination of two harmonic vibrations acting at a right angle has also been accomplished mechanically and can be demonstrated to the eye in the compound pendulum of

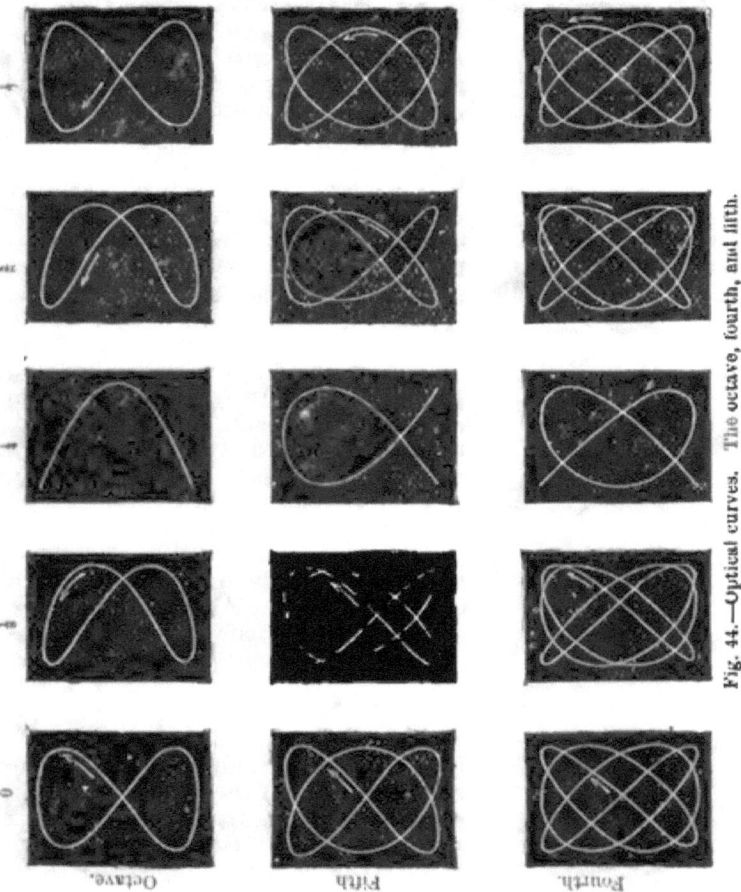

Fig. 44.—Optical curves. The octave, fourth, and fifth.

Professor Blackburn, to which Mr. Hubert Airy attached a tracing pen; thus rendering the figures permanent. It has been carried to still greater perfection in the instruments of Messrs. Donkin and Tisley. It was also in great measure accomplished by the Kaleidophone of Wheatstone, in which a silver bead attached to the end of a rod produced, when set in vibration, the patterns engraved above by virtue of the momentary persistence of retinal impressions. (2) It is also demonstrated by the "Vibration Microscope" of Helmholtz mentioned elsewhere.

(3) **Kœnig's Manometric Flames.**—This method consists in transmitting the movements of sonorous waves through a thin membrane of caoutchouc to a small reservoir of ordinary gas connected with a flame. A capsule of wood or metal is divided in the middle by such a membrane, forming two compartments. One of these is continuous with the vibrating mass of air, the other with the ordinary gas mains, and with a burner (Fig. 45). The membrane thus forming part of the wall of the pipe, yields to the alternate condensation and rarefaction of the air, and transmits these alterations of pressure to the stream of gas. The result is that the flame flickers up and down in coincidence with the vibration to be observed (Fig. 46). To render the flickerings distinct from one another they are received on a rotating mirror. While the flame burns steadily, there is thus formed a continuous band of light. But if the capsule is connected with a tube sounding its foundation notes the flame takes the form shown in the first figure (Fig. 47). If the octave be sounded it assumes that in the second.

If two tubes simultaneously give the fundamental and its octave, we obtain the following appearance (Fig. 48). If the interval be that of a third, the flame takes the more complicated shape shown below (Fig. 49).

Fig. 45.—Open tube with manometric flame.

Clarke and MacLeod's Method.—A new method for investigating and determining velocities of rotation has recently been published in a paper read before the Royal Society in April 1877, which besides its

IV.]   PITCH.—ITS MEASUREMENT, ETC.   95

other useful applications, affords an excellent, probably the best, method of determining the period of tuning-forks. It

Fig. 46.—Apparatus for the comparison of the vibratory movements of two sonorous tubes.

is founded on the following principles, and is termed the Cycloscope.

96   ON SOUND.   [CHAP.

" If a circle of dots equally spaced rotates in front of a tuning-fork provided with a lens or mirror; then if the fork is so arranged that it imparts to the image of the dot a

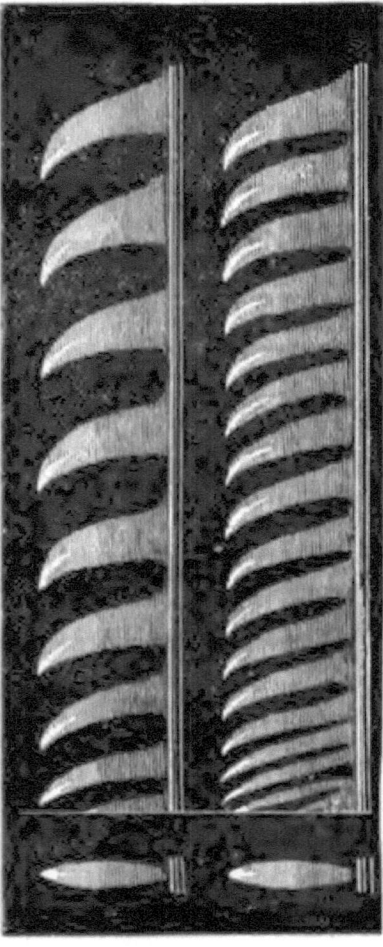

Fig. 47.—Manometric flames. Fundamental note, and the octave above the fundamental note.

movement at right angles to the motion of the latter, this double movement aided by the continuity of vision will produce on the eye the impression of a wavy line. The form of this wave line will depend on the ratio of the number of vibrations of a fork in any given time, to the number of dots which pass before it in the same time. With certain simple ratios, waves are produced whose forms are easily recognisable.

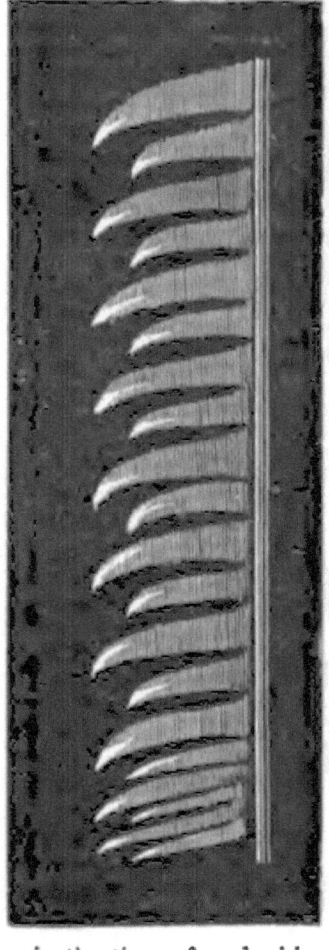

Fig. 48.—Manometric flames simultaneously given by two tubes at the octave.

If one dot passes in the time of a double vibration, a single wave is produced. If two dots pass in the same time, a double wave is produced. The double figure seems the best suited for measuring velocities.

H

"Now if the *exact* ratios stated above obtain, the waves corresponding to them will appear to be stationary. If however the speed of rotation is a little too quick for the exact ratio,

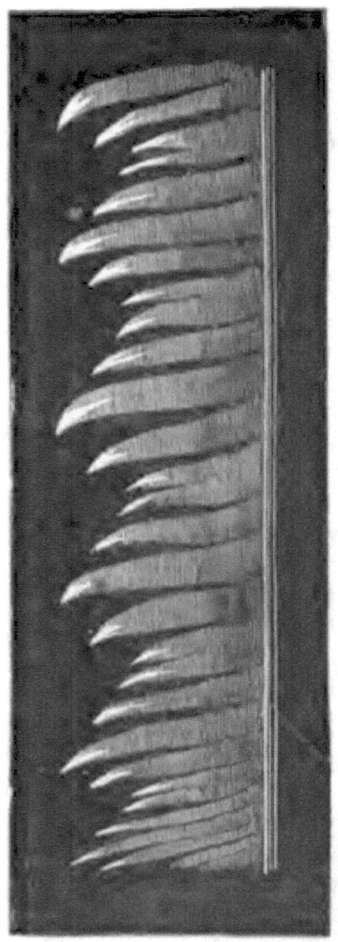

Fig. 49.—Manometric flames of two tubes of a third.

the wave will have a slow progressive motion in the same direction as that of the dots, while if the speed of rotation is a little too slow, the wave will move slowly in the opposite direction."[1]

[1] Extracted from a pamphlet by the inventors. See also *Proceedings of the Royal Society* for April, 1877.

IV.] PITCH.—ITS MEASUREMENT, ETC. 99

It is not necessary to enter into all the details of this valuable contrivance; but it is obvious that as with a tuning-fork vibrating at a standard rate, velocities of rotation can be accurately determined, so with a steady standard of rotation, the error of a tuning-fork from its theoretical vibration-number can be immediately detected.

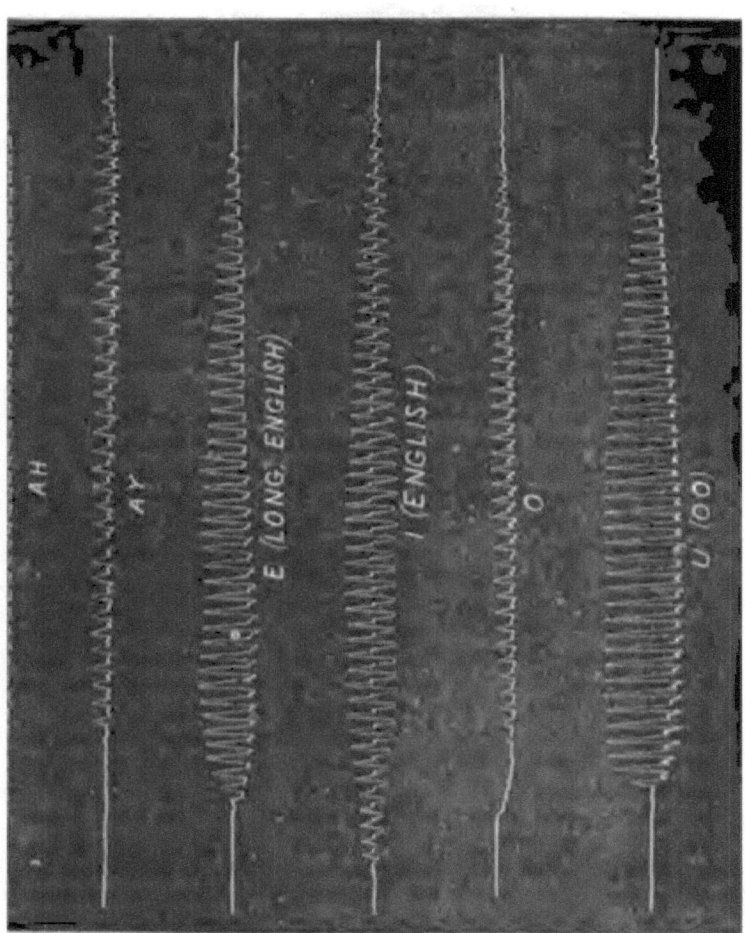

III. (1) **Photographic Methods.**—Professor Blake of Broner University has recently contributed to the *American Journal of Science* an ingenious method of photographing vibrations.

From the centre of the vibrating disc, made of thin iron as in the telephone, a wire projects which is connected by a short arm with the back of a small steel mirror capable of rotating in a vertical direction between two steel points. The reflecting surface of the mirror is firmly fixed perpendicular to the vibrating disc. A heliostat sends a beam of sunlight horizontally into a dark closet, and at a distance of several feet falls upon the mirror, which is inclined 45° to the horizon. The rays reflected vertically downwards pass through a lens, at the focus of which they form a luminous image of the opening of entry. A carriage moving smoothly on four wheels travels beneath the lens at such a distance that a sensitized plate laid upon it is at the focus for actinic rays. Uniform velocity is given to the carriage and is measured by a tuning-fork of 512 vibrations fitted with a style. If the carriage be set in motion alone a straight line is photographed. But on causing the disc to vibrate, each of its movements carries the reflected beam from the oscillating mirror through twice the angle of the mirror's motion. Curves are thus recorded on the photographic plate, the abscissæ of which are measured by the known velocity of the plate and carriage, and which serve to determine the pitch; the ordinates representing the amplitude of vibration of the centre of the disc magnified in this case 200 times. With the voice, speaking in an ordinary tone an amplitude of nearly an inch is obtained. This contrivance has been applied more to the analysis of vowel-sounds than to determinations of pitch; though it is obviously a form of graphic determination, and therefore deserving of record in this place.

The beautiful transcript on p. 99 was obtained by this method.

(2) **Scheibler's Method.**—The first person who hit upon a practical method of obtaining exact measurement was Scheibler, of Crefeld, who described it in a pamphlet published in 1834. His system, as modified by Helmholtz's more recent researches, is thus described.[1] If we strike a tuning-fork, with another an octave higher, and hold them both over their proper resonance chambers, we shall hear no beat, even if they are out of tune. But if they are applied to a sounding-board, a beat may be heard between the second partial of the lower, and the lowest of the upper fork. If both be held over the resonance chamber of the upper fork, the beat is heard more distinctly. Counting the number of beats in a second, they give the difference between the number of vibrations of the upper fork, and double the number of the lower. Suppose

---

[1] *Journal of Society of Arts*, May 1877.

the forks P and Q, of which Q is nearly an octave above P, when sounded together beat n times in a second, where n is less than A; then n will be the difference between the number of vibrations in Q, and twice those in P. But the beats do not tell us whether Q is too sharp or too flat. To discover this, tune a third fork R sharper than Q, and making four beats in a second with it. Try the beats of P and R. If they are $n + A$, then Q was too flat; but if they are $4 - n$, then Q was too sharp. Using P, Q, R, to represent the pitches of these notes, we have, when Q is too flat, $2P = Q + n$, and when Q is too sharp, $2P = Q - n$. Then interpose between P and Q (supposed to be too flat) so many tuning-forks that each beats with its neighbour, or with P or Q, either four times in a second or less. Suppose the sum of these beats to be m. Then $Q = P + m$, and since $2P = Q + n$, we have $P = m + n$; so that the number of vibrations is the sum of all the beats heard, including those with the imperfect octave. But when P is known, the pitch of all the intermediate forks is also known. In Scheibler's case he tuned Q so as to make no beats with $2P$; so that $Q = 2P$, $n = o$, and $P = m$. He then constructed 52 forks, and found $P = 219\frac{2}{3}$, $Q = 439\frac{1}{3}$, and the pitch of the intermediate forks was regulated so as to include the complete equally-tempered scale. In this case P was A on the second ledger-line below the treble staff, and Q was the usual A on the open string of a violin. Scheibler subsequently adopted 440 in place of $439\frac{1}{3}$ for his A. Having this scale complete, he had only to sound another fork beside these in turn to determine, by counting the beats, between which two it lay, and by how many vibrations it was sharper than one, and flatter than the other, the exact pitch of which was known. Very slow beats are difficult of observation, from the difficulty of distinguishing the falling off of the sound which naturally occurs from that which is due to interference. Scheibler, therefore, somewhat modified his plan, by using forks which were intentionally dissonant with the note to be determined, and beating with it some convenient number of times in a second. By reproducing the same number of beats on the instrument to be tuned as existed between the original standard and the intermediate fork, great accuracy could be obtained.

A still closer approximation has been obtained in counting beats, by the employment of a metronome pendulum, the period of vibration of which can be varied, and against which the beats may be compared. This instrument was originally used by Scheibler, but has been materially improved by Mr. Bosanquet.

(3) **Appunn's Reed Tonometer.**—Appunn has substituted free harmonium reeds for Scheibler's forks. These, although somewhat more affected by changes of temperature than tuning-forks, have the advantage of producing a louder and more coercive tone. The reed quality is peculiarly rich in upper partials, producing a strong, hard, and somewhat harsh sound, which is of great service scientifically, because it perfectly discriminates all the consonances, allowing a slight error to be immediately detected by dissident beats. The arrangement of Appunn's instrument is as follows:—Sixty-five reeds are arranged in a long rectangular box, and excited by a steady wind-pressure. The reeds each act in a separate chamber, controlled by a wire which opens a valve fully, or to any smaller amount. By pushing in the valve, the note is flattened up to about $2\frac{1}{2}$ vibrations in a second. The reeds are so tuned that each beats exactly four times a second with either of the adjacent reeds. The lowest is numbered 0, and the highest 64. Consequently the highest is four times 64, or 256 vibrations sharper than the lowest. The lowest and highest, sounded together, make a perfect octave. The difference between the numbers of vibrations being 256, it follows from what has been shown above, that the lowest reed makes 256, and the highest 512 vibrations in a second.

Unfortunately this apparatus is materially influenced by the power which the reeds, when vibrating strongly, have of influencing one another. Determinations made with it by Mr. Ellis are disputed by M. Kœnig, and Lord Rayleigh has recently added some excellent evidence to the same effect.

IV. (1) **Mayer's Electrical Tonometer** is described by Mr. Ellis, from an unpublished letter of the inventor. The seconds pendulum of a clock has a wedge of platinum foil attached to its lower extremity, which, at every swing, passes through a globule of mercury, placed vertically under it in the cup of an iron binding-screw connected with one wire of a small battery, of which the other wire is connected to the primary coil of a large inductorium, whence a wire passes to the top of the pendulum. By two other wires the secondary coil of the inductorium is connected with a tuning-fork and a revolving cylinder. The tuning-fork carries a delicate piece of platinum, which, as the cylinder revolves, will mark a curve on its smoked surface. Every time the pendulum leaves the mercury globule, a single spark is projected from the foil on the fork, which pierces the covering of the cylinder, and marks the beginning and end of the second. As the mercury globule may not be truly under the point of suspension, the length of every two seconds is used. The number of sinuosities of the curve between the spark holes, divided by two gives the pitch of the fork. This method is very accurate, but seems to be slightly influenced by the weight of the platinum on the fork, and also by the friction on the cylinder, as will be noticed further on.

(2) **Lord Rayleigh's Experiment.**—[1] A standard fork by Kœnig which was supposed to give 128 vibrations in a second was excited by means of a bow, and the object was to compare its frequency with the seconds pendulum of a clock keeping good time. The remainder of the apparatus consisted of an electrically maintained fork interruptor with adjustable weights, making about $12\frac{1}{2}$ vibrations per second, and a dependent fork, the frequency of which was about 125. The current from a Grove cell was rendered intermittent by the interruption, and as in Helmholtz's vowel experiments excited the vibrations of the second fork, the period of which was as nearly as possible an exact submultiple of its own. When the apparatus was in steady operation, the sound emitted from a resonator associated with the higher fork had a frequency determined by that of the interruptor and not by that of the higher fork itself; nevertheless an accurate tuning now necessary in order to obtain vibrations of sufficient intensity. This tuning was effected by prolonging as much as possible the period of the beat heard when the dependent fork starts from rest. The beat may be regarded as due to an interference of the forced and natural notes. By counting the beats during a minute of time it was easy to compare the higher fork and the standard with the necessary accuracy; all that remained being to compare the frequencies of the interruptor and of the pendulum. For this purpose the prongs of the interruptor are provided with small plates of tin so arranged as to afford an intermittent view of a small silvered bead carried by the pendulum, and suitably lighted. Under the actual circumstances of the experiment, the bright point of light is visible in general in twenty-five positions, which would remain fixed if the frequency of the interruptor were exactly twenty-five times that of the pendulum. In accordance, however, with a well known principle, these twenty-five positions are not easily observed when the pendulum is simply looked at; for the motion then appears to be continuous. The difficulty is easily evaded by the interposition of a somewhat narrow vertical slit through which only one of the twenty-five positions is visible. In practice

[1] *Nature*, November 1, 1877.

it is not necessary to adjust the slit to any particular position, since a slight departure from exactness in the ratio of frequencies brings all the visible positions into the field of view in turn.

In making an experiment the interruptor was tuned at first by sliding the weights, and afterwards with soft wax, until the interval between successive appearances of the bright spots is sufficiently long to be conveniently observed. With a slow pendulum there is no difficulty in distinguishing in which direction it is vibrating at the moment when the spot appears on the slit, and it is best to attend only to those appearances which correspond to one direction of the pendulum's motion. This will be best understood by considering the case of a conical pendulum, whose motion, really circular, appears to an eye situated in the plane of motion, to be rectilinear. The restriction named then amounts to supposing the hinder half of the circular path to be invisible. On this understanding, the interval between successive appearances is the time required by the fork to gain or lose one complete vibration as compared with the pendulum. Whether the difference is a loss or gain is easily determined in any particular case by observing whether the apparent motion of the spot across the slit, which should have a visible breadth, is in the same or the opposite direction to that of the pendulum's motion. The interruptor gained one vibration on the clock in about 80 seconds, so that the frequency of the fork was a thousandth part greater than 12·5 or 12·51. The dependent fork gave the ninth harmonic, with a frequency of 125·1. The beats between this fork and the standard, the pitch of which was the higher, were 180 in sixty seconds, so that the frequency of the standard was as nearly as possible 128·1 agreeing very closely with Kœnig's scale. The error of the determination may have amounted to ·1 but could not well exceed ·2. The approximate determination of the frequency of the interruptor had to be made independently, as the observation on the pendulum does not decide which multiple of $\frac{1}{2}$ most nearly coincides with the frequency of the fork. Also the relation between the two auxiliary forks was assumed; but on this point there could be no doubt, unless it be supposed that Kœnig's scale is in error to the extent of a whole tone.

V. (1) **Chladni's Rod Tonometer.**—It has already been stated that the vibrations of rods with one end fixed are inversely proportional to the square of their length independent of their area or cross-section, and that they give a series of harmonics following the squares of the odd natural numbers. It is not therefore surprising that they should have been used in the construction of the first computative tonometer by Chladni. This fact requires mention in the present chapter so as to complete the tabular statement. It is obvious however that tuning-forks, which have been shown to involve a rather different law of rod-vibration, are obviously preferable, and on them is founded the next, and best of all proposed methods.

Standards of Pitch.—It will appear from the foregoing observations that the absolute determination of the number of vibrations in a given period is a matter of considerable difficulty. The earlier mechanical methods furnish only rough approximations. It is, however, equally obvious that the necessity for such a determination, and for trustworthy standards which when once made may be kept for reference, is most urgent. The method of Scheibler, says Lord Rayleigh, " is laborious ; but it is probably the most accurate for the original determination of pitch, as it is liable to no errors but such as care and repetition will eliminate. It may be observed that the essential thing is the measurement of the *difference* of frequencies for two notes, whose *ratio* of frequency is independently known." What, then, is the best standard of reference ? There can be no doubt that, for the quiet of the physical laboratory, nothing equals a well-constructed set of Scheibler's tuning-forks.

But for practical purposes of music, the case is far different. The tone of a tuning-fork is too feeble and evanescent for the bustle and noise of a concert room. The quality of the note, moreover, is so pure and simple, that the beats produced are of very moderate intensity. In all these respects the free reed of the harmonium is far more marked and coercive. It will be shown that, although the variation with change of temperature is somewhat greater than in the case of a fork, it is still very much smaller than that of the organ, and it sinks into insignificance beside that which occurs in stringed or wind-instruments.

Variation of Standard Pitch.—The best testimony to be obtained on this subject is from organs and authentic tuning-forks. Mr. Ellis, in a laborious paper read before the Society of Arts, has collected a very large number of examples, and has "compiled a complete scale of variation of standard pitch proceeding from Handel to the present day, with three older and much lower isolated pitches." The following is an analysis of his researches. It may be noticed that his initial determination of the French normal pitch by means of Appunn's tonometer is still open to some doubt, on account of discrepancies due to temperature and the mutual influence of the reeds on one another. He states that instead of 435 or 870 the normal A really has a vibration-number of 439 or 878. This statement has been attacked by M. Rudolph Kœnig. But there is no reason to suppose that the relative pitches of the various standards examined suffer from this initial difficulty which may be easily remedied by a small constant correction to be afterwards applied.

Mr. Ellis makes five principal groups, as follows:—

I. *Ancient Low Pitch*, C below 500.—The first thing that strikes us is the great flatness of the older pitches. Dr. R. Smith's D, 262, in 1755, taken an octave higher, as D 524, gives nearly the present French normal for C. Hence his pitch was almost exactly a whole tone flatter than the present French C, and a tone and a quarter flatter than Broadwood's present high pitch, which we may take to represent "concert pitch." But rejecting this as in all probability wrongly ascertained, we have the fork tuned to Father Schmidt's C pipe at Hampton Court before the organ was reconstructed, and this is more than a semitone flatter than C 512, and $\frac{7}{8}$ of a tone flatter than "concert pitch." It follows that vocal music composed a hundred years ago ought to be transposed a whole tone, if sung at the present pitch, to produce its proper effect.

II. *The Handel Pitch*, C 500 *to* 513.—C 512 which was insisted on so strongly by Sir John Herschel at the meeting of the Society of Arts, to consider the Report of the Committee on Pitch in 1860, was in favour 50 to 100 years ago. Wieprecht gives a Berlin pitch of that amount, but the measurement may be doubted. We find, however, the fork to which Mr. Peppercorn tuned pianos for the Philharmonic concerts in 1815 was about 511, and reckoning by the old tuning, a fork used at the Plymouth Theatre about 1800, gives nearly the same, while Handel's fork of 1751 gave C = 510. We may take then the dawn of modern pitch to be C 512, which would be fully a semitone flatter than the present concert pitch. Hence vocal music of Handel's time should be transposed a semitone lower than it is written when played at concert pitch. The same remark applies to the music of Mozart, and probably of Haydn and Beethoven.

The value of C 512 which appears to have been aimed at about the period of this group, is entirely arithmetical. It has no other particular advantage. Arithmeticians can deal with any other C with equal ease by means of decimals. In measuring pitch it is never necessary to consider more than two places of decimals, and even the last place is used only to prevent an accumulation of error.

III. *French Normal Pitch*, C 514 *to* 527.—About forty years ago there was a French pitch in use almost coincident with that theoretically established in Paris in 1859; one fork from a good maker measured by Scheibler in 1834 actually gave A 434·9 or practically = C 517. The pitch, however, must have risen rapidly to about A 452, and the object of the French Commission was to regain this older pitch. This modern version of the older fork in the Paris Opera and Conservatoire was preceded in England by the almost identical but flatter pitch of Sir George Smart, and Broadwood's vocal pitch, and also by the very slightly sharper pitch of Scheibler in Germany, which being chosen by him as the mean pitch of Vienna grand pianofortes, represents the Vienna pitch of the time. From having been accepted by a congress of German physicists, who met at Stuttgart in 1834, it is commonly known as the Stuttgart pitch. Altogether, this group, which is comprised within about a quarter of a tone, represents that most in vogue now on the Continent, and consequently has the greatest claims on our attention, although its highest forks are $\frac{3}{5}$ of a semitone below our present high pitch.

IV. *Medium Pitch*, C 520 *to* 536.—The interval of about $\frac{1}{3}$ of a tone between the French normal and high pitch is not well marked. We have indeed within this group a fork from Leipzig, purporting to be the Dresden low pitch; one from Vienna, measured by Scheibler, but differing materially from the other Vienna forks; one from the Liceo Musicale at Bologna, in 1869; the medium pitch empirically adopted by Messrs. Broadwood and in the organ of St. Paul's. There were also several foreign forks in this group. The theoretical fork of the Society of Arts which begins it, was never really made, and Griesbach's A, like Hullah's C, were accidental errors. It would seem that the whole of this group is not generally satisfying; it is both too sharp and too flat, and can only be regarded as a neutral medium pitch.

V. *Modern High Pitch*, C *above* 536.—The highest group contains the moderately high pitch which the French Commission found so excessive, and the still sharper English concert and military pitch of the present day, with the high pitch of Brussels, strongly advocated in a report of a committee to the Belgian Minister in 1863.[1] We find that there

[1] The pitches given should be corrected by subtracting 4 for the error which Mr. Ellis attributes to the French Normal.

was a tuning-fork in Paris in 1826 giving A 445 = C 529·2 for the French Opera, another giving A 449·5 = C 534·6 for the Italian Opera, and another A 452 = C 537·5 for the Opera Comique. The two first belong to the preceding group. Many operas were composed to the last pitch which was afterwards raised to A 455 = C 541. The report mentions that when the French Commission was appointed the Opera pitch was A 453 = C 538·7, and that Lissajous wished to lower it to A 449·5 = C 534·6, but that a contrary opinion prevailed. The Committee say that to this high pitch belong the tuning-fork of the Brussels Conservatoire, one in use at Ghent, an old fork of the Paris Opera Comique in 1820, the tuning-fork of the Philharmonic Society of London, that of the Berlin Opera in 1861, and lastly that of the Choral Society of Cologne.

We have thus, according to Mr. Ellis's observations, a rise for C from 467 to 546, or 80 vibrations = 2·6 semitones in 130 years, or, if the early observations be rejected as possibly erroneous, from the undoubtedly authentic fork of Handel, which gives 507·4 to the vibration number of 546·5, at which the band of the Belgian Guides were playing in 1859. The writer can state from his own careful observations made at the Handel Festival of 1877, during the performance of the Israel in Egypt, on an extremely hot day in June, the thermometer being nearly 80° under the dome of the orchestra, that the pitch of A rose to 460, which is equivalent to a C of 547, and is higher than any previously recorded.

Causes of the Rise in Standard Pitch.—It will be seen that the tendency of the standard pitch has always been to rise, except when authoritatively and suddenly lowered, as in France, and more recently, though with little success, in England. It is also obvious that the rise occurs chiefly in orchestral performances. Much of this is due to temperature. All wind instruments rise with the warm breath of the player, especially the clarinet, which varies almost a semitone. Some part is also due to the fact that stringed instruments tune to perfect fifths, which can be shown to be incommensurable, from their larger interval, with the octave. Considerable weight must be given to the fact that the ear is physiologically liable to select the sharper of two notes for imitation; but the chief cause is an instinctive but vulgar inclination in the players themselves to give their own instrument an undue prominence at the expense of the others by slight sharpening.

Alteration of Pitch from Motion of Source.— It is clear that the pitch of a sound is liable to modification when the source and recipient are in relative motion. This fact was first stated by Doppler, and has been experimentally verified by Buijs Ballot and Scott Russell, who examined the alterations of pitch of musical instruments carried on locomotives.

It is clear that an observer approaching a fixed source will meet the waves with a frequency exceeding that proper to the sound by the number of wave-lengths passed over in a second of time. If v be the velocity of the observer and a that of sound, the frequency is altered in the ratio

$$a \pm v : a$$

according as the motion is towards or from the source. Since the alteration of pitch is constant, a musical performance would still be heard in tune, though when a and v are nearly equal the fall in pitch would be so great as to destroy all musical character. If we could suppose v to be greater than a, sounds produced after the motion had begun would never reach the observer, but sounds previously excited would be gradually overtaken and heard in the reverse of the natural order. If $v = 2a$ the observer would hear a musical piece in correct time and tune, but backwards.[1]

Similar results occur when the source is in motion and the observer fixed. With a relative motion of 40 miles per hour the alteration of pitch amounts to about a semitone. This can easily be substantiated by watching the whistle of a locomotive passing through a station.

A laboratory instrument for demonstrating this phenomenon has been invented by Mach, consisting of a tube six feet long turning on an axis at its centre; at one end is a whistle or reed blown by wind forced through the axis; when this is made to rotate rapidly, the pitch is found to fluctuate by an observer standing at the side, according as the rotating arm is approaching or moving away from him.

Kœnig uses two C tuning-forks, giving four beats with one another. If the graver of them be made to approach the ear while the other remains at rest, one beat is lost for each two feet of approach; if however the more acute be moved, one beat is gained by the movement.

[1] Rayleigh *op. cit.* II, 140.

CHAPTER V.

NATURE OF MUSICAL TONE. QUALITY. HARMONICS. RESULTANT TONES.

Nature of Musical Sounds. Quality.—It has been accepted as an axiom that the sensation of musical tone is due to a rapid periodic motion of the sonorous body; that of noise, to non-periodic motions; and it has been shown that musical tones are distinguished: 1, By their force or loudness. 2, By their pitch or relative height. 3, By their quality.

Quality.—It is to the third of these constituents that attention is now to be directed. The *quality* of a tone, which was formerly denoted by the anglicised French term *Timbre*, is that peculiarity which distinguishes the violin from the flute or the clarinet, and these from the human voice, when uttering sounds of the same pitch or frequency. Until the researches of Helmholtz this last characteristic had remained unexplained. He showed in a conclusive manner that the observed differences depended on no abstruse or recondite property, but simply on the co-existence with the principal of other secondary and affiliated vibrations, which accompany and modify the sensation by alterations which they produce upon the form of the sound-wave itself. Helmholtz aptly illustrates the possibility of difference even in periodic motions such as are so slowly performed as to be capable of being followed by the unassisted eye. For instance, the motion of a pendulum or an ordinary vibrating spring is one which is rapid in the middle of its path, and slow at either extremity: that of a hammer moved by machinery is marked by being slowly raised and falling suddenly. A ball thrown up vertically and caught on its descent by a blow which sends it up again to the same height, occupies the same time in rising as in falling, but at the lowest point its motion is suddenly interrupted; whereas above it passes through gradually diminishing speed of ascent into a gradually increasing speed of descent. None of these forms of motion are similar to one another, nor would they, if translated into sound, produce the same effect. The pendulum may be graphically represented by the double curve of sines before named, passing equally on either side of a straight middle line; the hammer by a series of long inclined planes terminated by a short downward curve; the ball by a series of arches abruptly reflected from one side of a base line. It is upon this difference in the form of the vibration that quality of tone depends. Ohm was the first to declare that there is only one form of vibration which will contain none of these secondary waves, and will therefore consist solely of the prime tone. This is the form peculiar to the pendulum and to tuning-forks, and hence they are called *simple* or *pendular* vibrations.

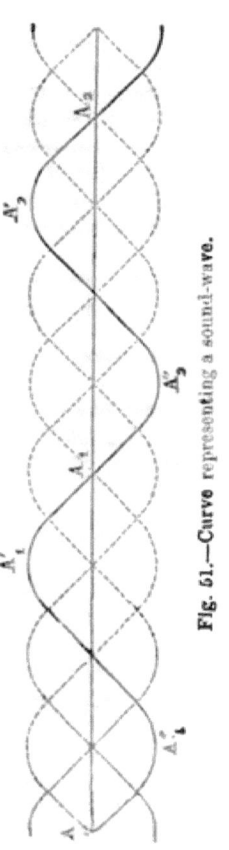

Fig. 51.—Curve representing a sound-wave.

Partials.—The affiliated or secondary waves, when occurring in the same period as the primary, are termed *harmonics*, *overtones*, or most accurately, *upper partial tones*. The character given to a particular note by their presence, by an over-literal translation of a German word, has been termed "clang-tint," it was formerly designated as timbre, but is best represented by the familiar English word quality.

When several resonant bodies simultaneously excite different systems of waves of sound, the changes of density of the air, and the displacement, and velocities of the particles of air within the ear are each equal to the algebraical sum of the corresponding changes of density, displacements, and velocities, which each system of waves would have separately produced if it had acted independently.[1]

The multiplicity of vibrational forms which can be thus produced by the composition of simple pendular vibrations is

---

[1] Helmholtz, *Sensations of Tone*, p. 43 *et seq.*

infinite. The French mathematician, Fourier, has proved the correctness of a mathematical law which may be thus enunciated : " Any given periodic form of vibration can always be produced by the addition of simple vibrations, having vibrational numbers which are once, twice, thrice, four times, &c., as great as the vibrational number of the given motion."

Fourier's Theorem.—For the purpose of application to the theory of sound, this law may be expressed as follows :—

Any vibrational motion of the air in the aural passages, corresponding to a musical tone, may be always, but for each case only in a single way, exhibited as the sum of a number of simple vibrational motions, corresponding to the partial tones of that musical tone.

It was, however, incumbent upon **Helmholtz** as the proposer of such a hypothetical explanation of quality, to show by experiment that the observed facts were in this way fairly represented. This he completely accomplished by the help of the resonators named above. Starting with an analysis of sympathetic vibration, he enunciates the law that "the simple partial tones contained in a composite mass of musical tones produce peculiar mechanical effects in nature altogether independent of the human ear and its sensations, and also altogether independent of merely theoretical considerations."

Commencing with the phenomena of reinforcement already given, sympathetic resonance is studied in a circular membrane strewed with sand, as in Chladni's investigations. It is shown that it is easiest to set the membrane in general motion by sounding its prime tone. The first form in which the resonator appeared was that of a pig's bladder stretched over the larger orifice of a glass receiver, the mass of contained air in the latter vibrating sympathetically with the membrane. It was found that such a membrane is not only set in vibration by musical tones of the same pitch as the proper tone of the bottle, but also by such musical tones as contain the proper tone of the membrane among their upper partial tones.

Resonators.—A great improvement, however, resulted from the substitution of the tympanum or drum of the ear itself for the membrane above described. Hollow spheres of glass or metal were made with two openings, one large with sharp edges, the other funnel-shaped and adapted for insertion into the ear. The globular or cylindrical cavity of the resonator is especially prone to vibrate in unison with its own prime tone, while it tends to damp all others. Hence any one, "even if he has no ear for music, or is quite unpractised in detecting musical sounds, is put in a position to pick out the required simple tone, even if comparatively faint, from a great number of others." Resonators of this kind are far more sensitive than membranes, and a series of them properly tuned is of great value in experiments where faint tones, accompanied by others of greater strength, have to be heard.

Analysis of Musical Tones.—It is, however, possible for the ear alone, unassisted by any apparatus, to resolve musical tones, in some degree, into their component partials. The analysis requires considerable practice, and at first is difficult. Success depends upon the power of concentrating the attention upon the sounds sought for, aided by proper methods for guiding the observer's ear towards the phenomena of which he is in search. It is advisable just before producing the musical tone which it is wished to analyse, to sound the note to be distinguished in it very gently, and, if possible, in the same quality of tone. The piano and harmonium answer well for such experiments, on account of the power of their upper partials.

In a monochord possessing a divided scale below the wire, Helmholtz recommends the following method for detecting the harmonics:—" Touch the node of the string with a camel's hair pencil, strike the note, and immediately remove the pencil from the string. If the pencil has been pressed firmly on the string, we either continue to hear the required partial as a harmonic, or else in addition hear the prime tone gently sounding with it. On repeating the excitement of the string, and continuing to press more and more lightly with the camel's-hair pencil, at last removing it entirely, the prime tone of the string will be heard more and more distinctly with the harmonic, till at last we have the full natural musical tone of the string. By this means we obtain a series of gradual transitional stages between the isolated partial and the compound tone, in which the first is readily retained by the ear."

The upper partials are less easy to hear on wind instruments and the human voice, than in stringed instruments, the harmonium, and the louder organ stops. Here the resonators become of great assistance, by directing the attention to the required tone.

Reality of Upper Partial Tones.—The proof that this analysis is real, and not imaginary, is obtained by exciting a string at any point in its length which is the node of a particular harmonic, in which case the harmonic in question will be absent. For instance, if it be plucked in the middle, the even partial tones will be absent, and the sound will have a peculiar hollow character. Or after it has been struck in the ordinary way, a camel's-hair pencil may be applied so as to damp all the simple vibrations which have no node at the point touched. A piano string plucked in the centre, and then touched at the same point, ceases entirely to sound, whereas, if plucked elsewhere, and touched in the middle, the second partial will be heard.

The result of these and similar experiments is that *the sensation of a musical tone is compounded out of the sensations of several simple tones.* The prime tone is generally louder than any of the upper partial tones, and hence it alone determines the pitch.

There are very few cases in which the sound of a resonant body consists of a simple tone, the chief instances being tuning-forks mounted on a consonant box, and large stopped organ-pipes blown with a very gentle stream of wind. On the other hand, the union of several comparatively simple tones into one compound of greater power is artificially produced in the organ by the stops termed cornets, sesquialteras, and mixtures. As a general rule—

1. The upper partial tones corresponding to the simple vibrations of a compound motion of the air are felt, even when they are not always consciously perceived.

2. They can be made objects of conscious perception without any other help than a proper direction of attention.

3. Even in the case of their not being separately perceived, because they fuse into the whole mass of musical sound, their existence in our sensation is established by an alteration in the quality of tone, the impression of their higher pitch being characteristically marked by increased brightness of quality, and apparently greater sharpness of pitch.

A list of the first and most important of these upper partials as founded on the bass C is as follows:—

The Synthesis of Tones.—The corresponding operation to the analysis of musical quality requires simple tones of great purity which can have their force and phase exactly regulated. These are obtained from tuning-forks, the lowest tone of which is reinforced by a resonance cavity, and thus communicated to the air. To set them in motion they are placed between the poles of an electro-magnet, the current in which is regulated by a separate apparatus. To make their sound audible, the resonance chamber has to be brought near them. Its mouth can be closed by a lid attached to a lever, by partial opening of which, the sound can be regulated to any amount of strength. The strings acting on a number of such levers are attached to a sort of key-board.

Helmholtz at first experimented with eight forks of this kind, representing $B\flat$ and its first seven harmonic partials, namely $b\flat, f, b'\flat\ d'', f''\ a''\flat$ and $b''\flat$. He afterwards had forks made of the pitch of $d''', f'''\ a'''\flat$ and $b'''\flat$.

To set them in motion, intermittent currents had to be passed through them equal in number to the vibrations of the lowest in a second, namely 120. The fundamental fork thus received an impulse at every vibration, the higher members of the series at periods inversely corresponding to their rapidity.

The moving apparatus consisted of a tuning-fork, to the two prongs of which were fixed platinum wires dipping in cups filled half with mercury and half with alcohol. These made contacts at every oscillation, which, sending an intermittent current through the electro-magnet, secured persistence of the vibration. A little steel clamp was placed on one prong of the fork, by which its oscillations might be brought into exact unison with the fundamental tone.

When the resonance chambers were closed, and the whole series of forks was in vibration, nothing was heard but a gentle humming sound, but on opening each of these, its corresponding tone was heard, and it became possible to form different combinations of the prime tone with one or more upper partials having different degrees of loudness, and thus to produce tones of different qualities.

The vowel sounds of the human voice are marked by comparatively low partials, those of E and I alone somewhat exceeding the limits of the forks used. U, O, the modified German $Ö$ and even A could be more, or less imitated. By further additions to the series of forks E, and indeed all but I were procured.

In the same way the quality of tone produced by organ-pipes of different stops was reproduced, also the reedy tone of the clarinet, by using a series of uneven partials, and the softer tones of the French horn by the full chorus of all the forks.

From all these experiments it was finally shown that the

quality of the musical portion of a compound **tone depends** solely on the number and relative strength **of its partial** simple tones, **and** in no respect on their difference of **phase.**

Vowel Tones.—If the lowest tone of the resonance chamber does not correspond with the prime tone, but with some of the upper partials, the corresponding upper partial is really more reinforced than the prime, and consequently predominates. The quality thus produced, more or less resembles the **vowels** of the human voice, which are really produced by the vocal chords, with the mouth as a resonance chamber, capable, by change of **shape and volume,** of reinforcing different partials of the **compound tone to** which it is applied.

With the assistance of resonators it is possible to recognise very high partials, **up to** the sixteenth when **one of** the more brilliant vowels is sung by a bass voice. Their loudness **differs** considerably in different persons, and they **are** generally more difficult to recognise by the unaided **ear than** in musical **instruments. The** investigation **of** the resonance **of the** cavity **of the** mouth **is** therefore of considerable importance. If tuning-forks of different pitches are **held** before it while **the** shape of the cavity is altered for the several vowels it **can be** determined which of them is most reinforced. **It can thus** be shown experimentally that the pitch of **strongest** resonance depends solely **on the** vowel for pronouncing **which** the mouth **has been arranged,** the resonances being essentially the **same in men, women,** and children.

The vowels **are arranged by** Helmholtz after **Du Bois** Reymond in this series :—

(1) The broad A as in *father* is the common origin of all, the cavity of the mouth **being** funnel-shaped, with uniform enlargement outwards. **A lower** series consists of O, as in *more,* and U **as** in *poor,* in which the lips are contracted and tongue depressed so as to form **a** bottle-shaped cavity.

(2) **In the** vowels A, E, I, and **the** modified German $\ddot{A}$ the lips are drawn **far** apart, and a **contraction** is made between the middle of **the tongue and the hard palate.** These have a higher and a deeper **resonance tone.**

(3) In the third **series from A through** O and U, in addition to the contraction of **the** tongue and palate, **we** have also a contraction of the lips into a sort of tube like that of a bottle with a very long neck.

The resonance of the cavity **of the** mouth for vowels may be thus expressed in musical notes :—

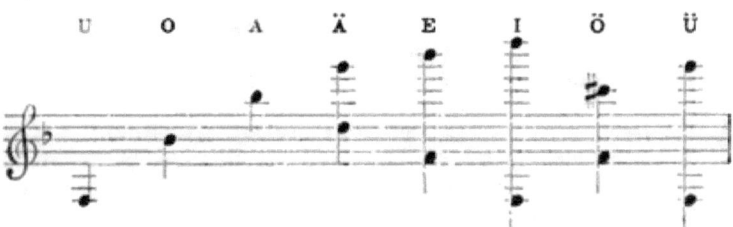

The theory of vowel sounds may be confirmed by means of artificial reed-pipes to which proper resonant chambers are added, as was done by Willis, who attached reeds to cylindrical pipes of varied length. The shortest gave *I*. Then *E, A, O,* up to *U*, until the tube exceeded the length of a quarter-wave. *E* and *I* however could not be well distinguished for the want of the second resonance. They can be better obtained with properly tuned resonators, such as glass spheres into the external opening of which small glass tubes from 2 to 4 inches long are inserted, so as to give a double reinforcement.

The Compound Nature of Musical Tone, as influencing the character or quality of sound, had long been known, and indeed followed of necessity from the fact that Quality or Timbre could not depend on frequency of vibration, which only influences pitch, nor on amplitude, which only affects loudness. It was first clearly shown by Helmholtz that "on exactly and carefully examining the effects produced on the ear by different forms of vibration we meet with a strange and unexpected phenomenon, long known indeed to individual musicians and physicists, but commonly regarded as a mere curiosity, its generality and great significance for all matters relating to musical tones not having been recognised." The discovery thus made, or at least brought into due prominence, was the law of Harmonic Upper Partial Tones. This fundamental principle of music may be accurately stated as follows:—

When any note is sounded on a musical instrument, in addition to the primary tone there are produced a number of other tones in a progressive series, each note of the series being of less intensity than the preceding. If m *denotes the number of vibrations of the primary tone in a given time, the vibration numbers of the whole series will be in the order* m, 2m, 3m, 4m, 5m, 6m, 7m, *&c.*

Thus, if the primary note be C, the whole series will be, for the first three octaves,—

Ratio	m	$2m$	$3m$	$4m$	$5m$	$6m$	$7m$
Note	C_1	C_2	G_2	C_3	E_3	G_3	X
Vibration number	64	128	192	256	320	384	448

We have here three C's, two G's, one E, and a note marked X lying between A_3 and B_3. If the octave above C be sounded with it, the harmonic series produced by C_2 will be, in the first three octaves,—

Ratio	$2m$	$4m$	$6m$	$8m$	$10m$	$12m$	$14m$
Note	C_2	C_3	G_3	C_4	E_4	G_4	X_2
Vibration number	128	256	384	512	640	768	896

again giving three C's, two G's, one E, and one X an octave above the former.

If then the two notes C_1 and C_2 be sounded together, the first three octave harmonics of C_1 will be compounded with the first two octaves of C_2, as follows :—

C_1 1st intensity.
$C_2 + C_2$ 1st and 2nd intensity.
G_2 3rd intensity.
C_3 2nd and 4th intensity.
$G_3 + G_3$ 3rd and 6th intensity.
E_3 5th intensity.
X_1 7th intensity.

It is therefore evident that when two notes C_1 and C_2 are sounded together, they produce *Overtones*, *Upper Partial Tones*, or *Harmonics*, in which, besides C's, the note G (the fifth) with its octaves, and the note E (the third) with its octaves, are more nearly related to C than any other notes are.[1]

Herein lies the physical explanation of the fact that "any sound and its octave bear the same name, in accordance with nature, since the two sounds so accord or tune together that they seem to be almost like one sound."

It may be remarked that the effect of the combination of notes differing by an octave is to throw the notes X further back, making them less audible, and bringing out more clearly the relationship of C, E, and G. Now this group of notes— which are actually sounded whenever C, either alone or with

---

[1] For further details see Haughton's *Natural Philosophy*, p. 170 *et seq.*, where this subject is treated with great lucidity.

any of its octaves is played—forms the Major Triad, the first, third, and fifth of the natural scale; their vibration numbers being in the simple ratio of the natural numbers 4, 5, 6.

The harmonic series has another most important bearing on music, which now may receive its full elucidation.

1. In the regular division of the sounding string,
2. By increasing the blast of wind in an organ-pipe,
3. Or by alteration of the embouchure in brass instruments,

the same order and sequence of sounds is obtained successively as has been here shown to co-exist simultaneously.

(1) The production of harmonics in stringed instruments has been already noticed. (2) In organ-pipes the use of harmonic stops, the consonant tubes of which are made twice their proper length, perforated with a small hole in the middle, and with this a high pressure of wind, illustrate the same principle. (3) In the French horn and similar instruments nearly the whole harmonic series is utilized in the scale of what are termed "open notes," as follows:—

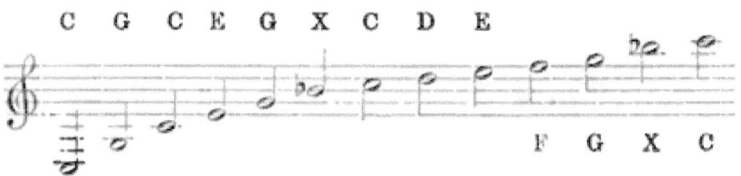

The real foundation note is of course an octave lower than the lowest here given; it is all but impossible to produce with the usual mouthpiece, but can easily be obtained by affixing to the tube some other source of sound, such as a clarinet reed.

If the harmonic series be extended to the full range of over five octaves, the seven sounds of the musical scale can be developed out of it in regular succession from the gradual approximation of the constituent ratios. This has been well demonstrated by Mr. Colin Brown, Euing Lecturer on Music at the Andersonian University, Glasgow, on the note F, where the twenty-fourth, twenty-seventh, thirtieth, thirty-second, thirty-sixth, fortieth, forty-fifth, and forty-eighth harmonics produce a correct enharmonic scale of eight consecutive notes from treble C to its octave. The harmonics which do not belong to the scale are marked with a cross on the approximate line of the staff to which they belong.

120　ON SOUND.　[CHAP.

Development of Scale from Harmonic Series.

It has been shown that C, whenever sounded, introduces E and G. In similar fashion it is possible, without using other notes than these and their octaves or natural representatives, to establish two other major triads, in this manner.

$$C : E : G :: 4 : 5 : 6 \quad \ldots \quad \text{Tonic triad.}$$
$$G : B : D_2 :: 4 : 5 : 6 \quad \ldots \quad \text{Dominant triad.}$$
$$F : A : C_2 :: 4 : 5 : 6 \quad \ldots \quad \text{Subdominant triad.}$$

B and D_2 standing as third and fifth to G; A and C_2 as third and fifth to F. For if G or F were sounded alone, or with octaves, they would respectively introduce B—D_2 and A—C_2.

The sensation of pleasure felt on sounding certain notes together depends therefore on the agreement of the harmonic sounds necessarily accompanying them, and on the simplicity of the succession of impulses produced on the ear by both primary and secondary tones. There is only one form of vibration which gives rise to no harmonic upper partial tones. This is peculiar to pendulums and tuning-forks, and may thence be called simple pendular vibration.

"A compound," says Helmholtz, "has, properly speaking, no single pitch, as it is made out of various partials having each its own pitch. By the pitch of a compound tone we mean the pitch of its lowest or prime tone. Every musical tone in which harmonic upper partial tones can be distinguished may be considered in itself as a chord or combination of various simple tones."[1]

Many of the sources of sound enumerated in an earlier chapter produce in conjunction with the prime tone, a number of inharmonic upper partials: such has been especially noted to be the case with bells.

Those which are commonly utilized for musical purposes, are characterized by harmonic upper partials. They are classified by Helmholtz according to their mode of excitement, namely: 1, By striking. 2, By bowing. 3, By blowing against a sharp edge. 4, By blowing against elastic tongues or vibrators. The two first classes comprehend stringed instruments alone; as rods vibrating longitudinally, (the only other instruments producing harmonic upper partial tones,) are not used for musical purposes. The third class embraces flutes and the flue stops of the organ; the fourth, all other wind-instruments, including the human voice.

[1] *Vide* Ellis's translation, p. 35.

Musical Tones of Strings.—The instruments of this class which are excited by plucking are the harp, guitar, and zither. In bowed instruments the same effect is produced and termed *Pizzicato*. The theory of their motion under these circumstances is complete. It is the same for a string which has been **struck in one point by a** hard sharp edge. In the pianoforte, where soft hammers are used, it is more complicated. The force of the upper partials depends on—1, The nature of the stroke. 2, The place struck. 3, The density, rigidity, and elasticity of the string.

(1) A **sharp point** produces **a shriller tone with a larger number of high upper** partials. When **the hammer is soft and elastic, the motion** has time **to** spread before **the hammer** rebounds, and greater softness **of** tone **results from the cor**responding decrease of the higher partials.

(2) As regards the place struck, if **it** approach the end of the string, the upper partials increase, whereas if it be nearer the middle, the even partials decrease and the sound is more hollow. The point struck in pianos is from one-seventh to **one-ninth of** the length from the end of the string, and then **the seventh and** ninth partials become comparatively weak. These are the **first in** the series which **do not** belong to the major chord **of the** prime **or** fundamental tone, and **their** exclusion renders **the** quality more **smooth** and pure.

(3) Rigid **strings** cannot form **very high** upper partials, and light strings **of** gut, on the **other hand, are** favourable to them; but the **inferior elasticity of this latter** material tends to damp them. **Hence plucked strings of catgut, as in** the guitar and **harp, are much less tinkling in sound than those of** metal.

Vibration Microscope.—The vibrations of **bowed strings were** examined by means **of the** vibration-miscroscope above named. They were found **for** the middle string of **a violin to** be essentially different **from** simple vibrations, **being** angular **and** sudden, like the **motion of the** tilt-hammer named previously. They resemble **the teeth of a saw, the** downward line being often **so** rapidly executed **as to become** invisible. During the greater **part** of each vibration, **the** string clings to the bow and is **carried on** by **it**; it then suddenly detaches itself and rebounds, being again **seized by** other points of the bow, and again carried forward.

Quality of Organ Pipes.—In the **flue** pipes of organs the quality differs materially with the diameter and length of the pipe itself. The tube which constitutes the air-chamber **of** the **pipe** strengthens by resonance such tones as correspond with its proper **tone,** and makes them predominate over the rest. It is only in narrow cylindrical pipes that **the** higher upper partials of the tube exactly correspond **with the** harmonic upper partials of the prime tone. **By** using **a** resonator, Helmholtz finds **that in** narrow pipes partial tones may be heard clearly up **to the sixth.** Wide pipes, on the contrary, produce the prime tone strongly and fully, with much weaker **secondaries.** The narrower stopped cylindrical pipes have proper tones corresponding to the uneven partials of the prime, the third partial or twelfth, the fifth partial or high major third, and so on. Wide stopped pipes when gently blown, give **the prime tone almost** alone.

Resultant or Combinational Tones.—Two **notes, when** sounded together, produce, under certain circumstances, other notes which are not actual constituents of either tone. These are termed resultant or combinational, and **are of two kinds,** differential and summational. The former **have a** vibration number equal to the difference of their components, the latter one which is the sum of their frequencies. The sounds com**bining to** produce this **result may** be either fundamental or upper partials; hence notes rich in harmonics may yield a large number of resultant tones. They were observed in the last century **by** Sorge **and** by Tartini, and were until lately attributed **to** beats, the frequency **of** beats having been shown to depend directly on the vibration-frequencies **of the tones** producing them; but this explanation fails to account for the phenomena, and would form an exception **to** the rule which appears to be general, namely, that every simple **tone arises** from a corresponding simple vibration. It does **not explain** the summation tones, and in certain cases the difference-**tones** and the beats can be heard together. The pitch of **a** combination-tone **is** generally different from that of either of those producing it or from that of their upper partials. Both primary and upper partial tones may give rise to **them.** The differential are generally stronger than the summation tones. As the prime tone generally predominates over the partials, the differentials of the former are more distinctly heard. To hear them, two sounds forming a just interval should be strongly held together; a weak low **tone** will be heard. **In** the fifth the tone thus produced is an octave below the lower generator; in the fourth a twelfth; in the major-third two octaves; **in** the minor third, two octaves and a major third; **in the major** sixth, **a** fifth; in the minor sixth, a major sixth, according to the following scheme:—

Difference Tones.

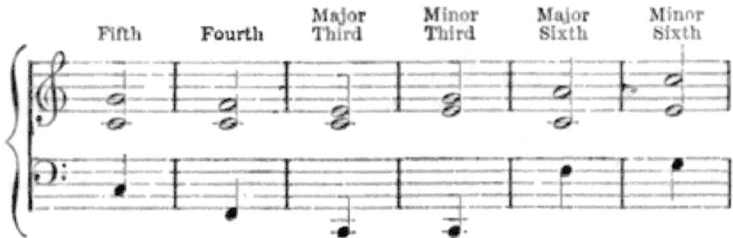

The summational tones are usually weaker than the differential. By their definition, they are always sharper than those producing them, as follows:—

Summation Tones.

In the last two cases the summation tone lies between the two notes given in the upper stave.

They are most easily heard on the polyphonic siren and on the harmonium when tuned to true intonation, of which indeed they may be used as a test. They come out very distinctly when two powerful soprano voices sing in thirds, and occasionally produce a very painful effect in the full chorus of a large organ.

Beats of Upper Partial Tones occur when the two generators are more than a minor third apart, and are of considerably more practical importance for musical purposes than the beats of combination tones. It will easily be seen that beats may occur whenever any two upper partials lie near together, or when a prime of one approaches an upper partial of the other. The number of beats is the difference of the vibrational numbers of the partials, or of the upper prime and the first partial of the lower. It is from the perfect agreement of the upper partials of notes standing truly in the relation to one another of the perfect octave, the perfect twelfth, and the perfect fifth, that these intervals are termed consonant; an agreement which had long ago been ascertained empirically by merely following the judgment of the ear respecting the pleasantness of the concord. On the other hand the beats of imperfect fifths have also been long known to organ-builders, who use them to obtain the required "tempering" of intervals. There is indeed no more sensitive means of proving the accuracy of intervals. To an unpractised ear it is often difficult to determine which part of the whole sound produces the beats, and here the resonators named above will be of great assistance. The following example of the coincidences of upper partials is given by Helmholtz.

When the differences are small, the beats are slow and easy to count. But as these become more numerous, their real character is concealed, and they give a rough quality to the whole mass of sound, which would not be recognised as the result of beats unless this had been step by step demonstrated.

The number of beats arising from putting one of the generating tones out of tune to the amount of one vibration a second is always given by the two numbers which mark the interval. The smaller number gives the beats which arise from increasing the vibrational number of the upper tone by 1. The larger gives those arising from increasing the lower by 1. If we sharpen A in the major sixth $C.A.$ by 1 vibration we have 3 beats; if we sharpen C a like quantity we have 5 beats in a second.

Beats may arise from combinational as well as harmonic upper partial tones; the loudest of these being the first differential tone. This is the chief agent in producing beats, the higher differentials and the summation tones being far weaker. The first differential tones, however, of compound notes cannot produce beats, except when the upper partials of the same generate them, and the rapidity of the beats is the same in both cases. Hence there is only a slight increase in the strength of the beats. But with simple tones the case is different, since with them no beats whatever could be produced unless they were nearly in unison. Such beats do nevertheless occur, though much weaker, even with wider intervals. In the imperfect octave the first differential is competent to produce beats, but in the fifth the beats are due to a more complicated relation, which increases in complexity for the fourth, and in the imperfect major third is hardly to be recognised.

Differentials and Upper Partials of Intervals.[1]

The notes of the interval are represented as minims, the differentials as crotchets; the second differentials to the left of the larger notes.

[1] From Curwen's *Musical Statics*.

CHAPTER VI.

EFFECTS OF HEAT, ATMOSPHERIC PRESSURE, MOISTURE, DENSITY.

Effects of Heat.—It has been shown by careful observations that the velocity of Sound in air at 0° Centigrade or freezing point of water is about 332 metres or 1090 feet per second. This increases with rise of temperature, being proportional to the square root of absolute temperature.[1] If t be the ordinary centigrade temperature, a the coefficient of expansion, ·00366, the velocity at any temperature may be found by the formula

$$1090 \sqrt{1 + a t} \text{ in feet per second.}$$

The velocity must be theoretically given by the formula

$$v = \sqrt{\frac{E}{D}}$$

where D is the density and E the elasticity of the air.

If P denote the pressure of the air in units of force per unit of area, and the temperature be kept constant during compression, a small additional pressure p will, by Boyle's law, produce a compression equal to $\frac{p}{P}$, and the value of E will be simply P.

On the other hand, if no heat is allowed to enter or escape,

---

[1] *Absolute Temperature* is the point at which the motion of gaseous molecules would cease. A gas is increased $\frac{1}{273}$ of its volume for each degree Centigrade, and at 273° is doubled. If the temperature of a given volume were lowered through 273°, the contraction would be equal to the volume; that is, the volume would not exist. In all probability, before reaching this temperature the gas would undergo a change of state.

the temperature will be raised by compression, and additional resistance will take place. The compression will therefore be $\frac{p}{P(1+\beta)}$ where $1 + \beta$ is the ratio of the two specific heats which for air and simple gases is 1·41. The value of E will be $P(1+\beta)$. Between these limits the velocity in air will vary. Observation shows it to be close upon the latter of these values.

Effects of Pressure.—Velocity is independent of the height of the barometer since pressure and density are affected to the same amount, and in the same direction. At Quito, where the mean pressure is only 21·8 inches, the velocity is the same as at the sea-level, provided the temperature be the same.

If g be the force of gravity, h the barometric height reduced to zero, and d the density of mercury at zero; then for a gas at atmospheric pressure $E = g h d$, and Newton's formula becomes

$$V = \sqrt{\frac{g h d}{d}}.$$

If temperature increases from $0°$ to $t°$ its volume will increase from unity at zero to $1 + a t$ at $t°$, a being the coefficient of expansion of the gas. But density varies inversely as the volume, therefore d becomes $d - (1 + a t)$ hence

$$V = \sqrt{\frac{g h}{d} (1 + a t)}.$$

Values of V thus obtained are however less than the experimental values, in consequence of the heat produced by compression, as already stated.

The coefficient of expansion of air, in a fractional form has been given as $\frac{1}{273}$ per degree Centigrade. Hence taking the velocity at freezing point, as 1090 feet per second, that at a given temperature $t°$ will be—

$$1090 \sqrt{1 + \frac{t}{273}}.$$

Converting this into degrees Fahrenheit it gives about 1110 ft. at 50° Fahrenheit and 1148 at 86° Fahrenheit, or an increase of velocity of about a foot per second for each degree Fahrenheit.

Influence of Moisture.—Aqueous vapour, being somewhat lighter than air, renders the density of the mixture somewhat less; about $\frac{1}{220}$ at the temperature of 50° Fahrenheit, and the corresponding increase of velocity is about 2 to 3 feet a second. On the other hand a heavy vapour like that of ether distinctly reduces the velocity, and lowers pitch.

Velocity in Gases.—These rules apply to gases generally. The velocity in any simple gas is

$$V = \sqrt{1\cdot 41 \, \frac{P}{D}}$$

D denoting density at pressure P.

In two gases at the same pressure, the velocity is inversely as the square roots of their absolute densities whether at the same or at different temperatures. **In liquids,** the changes of temperature from compression and extension are so small as to be neglected, and the same is to a certain extent true as regards solids. The change, however, of pitch in solid sources of Sound such as strings and tuning-forks, from alterations of temperature, depends on a totally different cause and requires separate consideration.

On Strings.—A vibrating string of metal, stretched between two fixed supports, materially alters its tension with heat from the expansion and contraction of the metal itself. There is also a modification of elasticity due to the same cause.

An experiment proving the latter proposition may be performed by extending a string of steel or platinum over a long trough of lighted alcohol. The string, fixed at one extremity, passes over a pulley, or on to the short arm of a bent lever weighted at the other end; and though the tension thus remains the same, the note sinks in pitch from the cause above named, and at last becomes extinguished.

The former influence is very perceptible in the pianoforte, which is materially sharper in frosty weather, the strings themselves not unfrequently snapping from this cause. The frame of the instrument, especially if made of wood, being less susceptible to variation of bulk from increments of heat, remains fixed, while the string shortens.

The writer has made experiments by passing an electric current through a steel and brass string strained on a sonometer. He found that the heat thus developed was competent to lower the pitch through the interval of an octave.[1] Even

---

[1] *Proceedings of Physical Society* and of *College of Organists.*

K

when the tension was produced by means of a weight, and the string thus allowed to lengthen, there was still a notable fall of pitch, due, in all probability, to alteration of its elasticity.

Strings of catgut, being very hygrometric, are also materially affected by moisture, which swells the material laterally and tends to shorten it. In a hot damp concert-room violins vary rapidly and somewhat irregularly from this cause.

On Tuning-forks.—The effect of heat on tuning-forks was noticed and roughly computed by Perronet Thompson. He states that they are made flatter by heat at the rate of a hundredth of a comma for every three degrees Centigrade (or 5·4 Fahrenheit). His experiment was conducted as follows. A steel tuning-fork sounding treble C was cooled to the freezing-point in snow and the load noted which brought the harmonic double octave of his monochord into unison with it; this was 240 lbs. The fork was then heated in boiling water; upon which 2 lbs. had to be taken off the load to bring it again into unison. The increase of length may be estimated at ·0147 of an inch to the foot, which is competent to flatten the pitch by less than the tenth of a comma. But the fork was found flatter by the *third* of a comma: more than two-thirds of the effect must therefore have proceeded from some other cause. "For which nothing so readily presents itself as a relaxation of the elastic power of the metal at the shoulder or bend. In confirmation of which it was observable that the forks, when heated to boiling-point, lost much of their strength of tone, and did not entirely recover it on cooling."[1]

Perronet Thompson's observations give the coefficient of alteration as ·000023. Scheibler gives as the results of his observations what is equivalent to ·0000573 at 440 vibrations and ·0000505 at 220 vibrations as the alteration of pitch per degree Fahrenheit. Professor MacLeod and the writer made the amount larger, namely ·000125 for 1° Centigrade or ·00007 for 1° Fahrenheit. Professor Mayer says that an ncrease or diminution of 1° Fahrenheit lengthens or shortens the fork by $\frac{1}{22000}$ part, thus making the coefficient of change ·00004545. The latter determination was made by exposing a fork in a room with an open window during four days of cold weather, noting the temperature, and another in a room heated to 70° by steam pipes. These discrepancies are doubtless due to the kind of steel employed for making the forks.

[1] *On Just Intonation*, p. 74.

Effect of Barometric Pressure.—When the barometer rises, the effect on strings, wires, and tuning-forks, is to increase the flattening attributable to the resistance of the air by $\frac{1}{10}$ for every inch of rise. From this cause the harmonic octaves are too sharp to the fundamental tone, an error which may be referred to the effects of the air's resistance on different lengths of string. To take the instance of the octave, the vibrations of this will be twice as many in a given time as those of the open string, but the extent of each vibration may be estimated as half; so that the whole space travelled over may be considered as unaltered. So far, therefore, the resistance of the air may be expected to produce the same effects on both. But there is another element, which is that the shortened lengths present proportionally less surface to the air. Hence the greater lengths will be more retarded than the smaller, and the increase of retardation of the longer strings, which is the same as over-sharpness of the shorter, may be expected to be as the differences in length. It is experimentally found to be so.

On Free Reeds.—In the harmonium reed the process of alteration by change of temperature is far more complicated. For whereas the reed itself is equivalent to a vibrating rod supported at one extremity, its vibrations slacken by its expansion, and by its diminution of elasticity; on the other hand, the air which it sets in motion becomes less dense, and transmits the sound-wave with increased velocity. The latter action predominates on the whole; hence free reeds do not flatten with heat as has been stated; but they sharpen so much less than flue-pipes in an organ as to produce the same effect. For this reason they are almost entirely disused in this combination. They are slightly inferior in this respect to tuning-forks, which, as above stated, alter, according to one of the higher estimates, ·0125 per cent. for 1° Centigrade, whereas from experiments recently made with a brass reed, it seemed to vary about ·0277 per cent. for 1° Centigrade of temperature. As standards of pitch, however, in spite of this trifling inferiority, they are far superior on account of their more incisive tone, the abundance of their upper partial tones, and the consequent loudness of the beats they produce.

On Organ Pipes.—The general effect of heat on an organ-pipe is to sharpen the note it emits. The compensatory effect named in the case of the reed occurs in this instance also, but to a much less extent. For the pipe itself lengthens, especially if it be made of a very expansile metal, such as tin or pewter. But the pitch of the flue-pipe is far more dependent on the contained column of air than on the envelope which surrounds it, and consequently the rarefaction of the medium tells upon the note with an influence quite predominant. In wooden pipes, the expansion of the tube is so very small as to be entirely inappreciable. Small pipes grow relatively sharper than large ones under the same increment of heat. Perronet Thompson states that the middle C pipe of the open diapason, two feet long, and two inches in diameter, sharpens a comma under a rise of 10° Centigrade, or 18° Fahrenheit, the barometer being stationary.

Different kinds of pipes, such as stopped and open diapasons, vary not only in the amount of sharpening, but also in the rate with which it takes place. Open pipes sharpen more quickly than stopped pipes, and metal pipes more rapidly than wooden ones. The smaller pipes are affected by changes of heat sooner than the large, as well as more.

In an organ of several stops, on suddenly raising the heat from 10° to 15° Centigrade (50° to 59° Fahrenheit), the smallest pipes of the metal open diapason grew sensibly too sharp for the others in the course of half an hour. This sharpening continued to increase for four hours, when they were too sharp for the largest pipes, by a quarter of a comma, the intermediate pipes in the meanwhile growing sharper, the smallest first. In the stopped diapason of wood, for four hours there was no perceptible alteration; but after that time the differences began to be sensible. During these processes, in consequence of smaller alterations in the wood, the open diapason sharpened upon the stopped; the greatest difference during the time of the experiment being at treble C, where the open diapason grew sharper than the stopped by three-eighths of a comma. In the enharmonic organ a rise in the thermometer of 10° Centigrade (18° Fahrenheit) raised the tuning C, termed the "master pipe," by a comma; a fall in the barometer of an inch, according to P. Thompson, did the same. He states generally that when the barometer and thermometer move the same way they act in opposition to each other, when they move different ways they act together. Stopped pipes are less affected by barometrical variations than open.

The writer has endeavoured to obtain some more accurate determinations on the subject of heat, as applied to reeds and organ-pipes in the following manner:—A wind-chest, capable of supplying a continuous stream of air at very equable pressure, is connected with two spirals of metal tubing, one of which is kept constantly at the freezing point of water by being immersed in melting ice, the other at boiling point by means of a cistern of boiling water. The issuing air is used to feed two similar pipes, both of which are compared with a standard at the ordinary temperature. Thermometers are inserted in the respective currents, and the variations of pitch studied by means of the beats. In all cases the rise and fall is gradual, much more so with wooden and stopped metal pipes than with open metal ones. It seems to depend chiefly on the warming of the metal walls of the pipe itself, which rapidly radiate back to the contained air. In the same manner a reed fixed in a small wind-chest, is alternately blown by means of air at 32° and 212° Fahrenheit. The changes can be noted against the standard named above. In the reed the alterations of pitch are considerably less than those of pipes, for the reasons given above. The best result seems to follow from the use of German silver vibrators, which alter their molecular condition very slightly with moderate increments of heat, as is well known to students of dynamic electricity.

CHAPTER VII.

SCALES, CHORDS, TEMPERAMENT, AND TUNING.

HITHERTO sounds have been treated as independent of one another, as bearing no mutual relations, and, except in the case of interference, as exercising no influence the one upon the other. This view represents only a limited, and what may be termed the physical side of acoustics. Beyond this lies the chief part, namely the æsthetic or musical conception of sound; which differs essentially from the former, in contemplating vibrations as intimately linked together, either in close series and succession, forming scales and melodies, or as simultaneously elicited, and furnishing the infinite varieties of chords and harmony. It is remarkable that whereas melody existed in ancient times, and has been cultivated by all nations, harmony in an extended sense, and possessing any pretension to exactness, is comparatively modern in its origin, and limited in its diffusion.

It was long known that the rapidity of vibration of a string under constant tension was inversely proportional to the length of the string, that is to say, that if we halve the length of the string, we double the number of its vibrations. To this we owe all power of playing on the violin, and also all knowledge of the relative pitch of the notes in the Greek and Arabic scales, for which the corresponding lengths of the string were given by Euclid the mathematician in the fourth century B.C. and by Abdul Kadir, the Persian theorist of the fourteenth century. Helmholtz points out that in the music of all nations so far as is known, *alterations of pitch in melodies take place by intervals, and not by continuous transitions*, and he further defines all melodies as *motions within extremes of pitch.*

VII.] SCALES, CHORDS, TEMPERAMENT, ETC.

It will be seen moreover, from the remarks on quality, that the scale, as now recognised, was potentially contained in all sounds from the first, and can be built up from the harmonics which accompany musical tones. For the sensation of interval between two notes is not due to the absolute difference but to the ratio of their pitches. In considering the scale therefore we enter on a new and independent subject from that of pitch, and casting aside the element of time, deal entirely with proportion.

In the case of the octave, the connexion with the fundamental tone is intimate: a melody sung on the human voice conveys to the hearer not only the primes of the compound tones, but also their upper octaves, and with less force, the other upper partials. Thus when a voice an octave higher, such as that of a woman or a boy, reproduces the same melody an octave higher we "*hear again a part of what we heard before,*"[1] nothing which we had not previously heard. The same is true in a less degree of the twelfth, though only a smaller part is repeated, and in giving the fifth we repeat the same tones with two new ones, the third and ninth.

This imperfect repetition of the fifth caused the Greeks to divide the octave into two tetrachords as follows:—

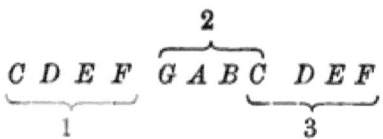

The second tetrachord is a reproduction of the first with the transposition of a fifth, the upper (or third) borrows a note from the second. Consecutive tetrachords must thus, in an octave scale, be successively separate and connected.

Intervals.—The discovery of the relation existing between the different notes of the scale dates back to Pythagoras, who divided a stretched string into three equal parts. On stopping the junction of these, the longer section was found to give the lower octave of the shorter. Then dividing the same string into two parts bearing the proportion of 2 to 3, he found an interval of a fifth between the two portions. In a similar way it may be shown that the ratio of 3 to 4 gives the fourth; that of 4 to 5 the major third, and that of 5 to 6 the minor third. By doubling the smaller number, which is equivalent to taking it an octave higher, other ratios

[1] Helmholtz, p. 390. Ellis's translation.

are obtained, namely, from 4 to 5 altered to 5 to 8, we have the minor sixth, from 5 to 6, changed to 6 to 10 = 3 to 5, the major sixth. These are all the consonant intervals which lie within an octave. Excepting the minor sixth, the most imperfect of them, all are expressed by the first six integers.

Hence the law first enunciated by Pythagoras, that *the simpler the ratio of the two parts into which the string is divided, the more perfect the harmony of the two sounds.* It was not, however, until much later, through the investigations of Galileo, Newton, Euler, and D. Bernoulli that the numerical foundation of the law was disclosed, and also the fact that these ratios governed all instruments, as well as the strings on which they had first been substantiated. The cause of this preference of the ear for simple ratios remained unexplained until quite recently. Even Euler was satisfied to believe that "the human mind had a peculiar pleasure in simple ratios" as being more easily comprehended. It was reserved for Helmholtz to show that the greater or less smoothness of the compound was due to the absence or presence of clashing upper partials, which have already been explained to accompany almost every fundamental tone.

Table of the principal Intervals, with their Ratios and Logarithms.

	Example.	Ratio.	Logarithm.
Schisma	♯B to C	$\frac{32768}{32805}$	·00049
Comma	♭C to C	$\frac{80}{81}$	·00539
Minor Semitone	♯C to C	$\frac{24}{25}$	·01773
Chromatic Semitone	C to C♯	$\frac{128}{135}$	·02312
Major Semitone	C to B	$\frac{15}{16}$	·02803
Minor Tone	E to D	$\frac{9}{10}$	·04576
Major Tone	D to C	$\frac{8}{9}$	·05115
Minor Third	G to E	$\frac{5}{6}$	·07918
Major Third	E to C	$\frac{4}{5}$	·09691
Perfect Fourth	F to C	$\frac{3}{4}$	·12494
Perfect Fifth	G to C	$\frac{2}{3}$	·17609
Minor Sixth	C to E	$\frac{5}{8}$	·20412
Major Sixth	A to C	$\frac{3}{5}$	·22185
Minor Seventh	F to G	$\frac{5}{9}$	·24988
Major Seventh	B to C	$\frac{8}{15}$	·27300
Octave	C to C	$\frac{1}{2}$	·30103

Chords consist of the simultaneous production of more than two **separate** compound tones. Some **of** these are termed concords, others discords. Concords can be formed by **taking two** consonant intervals to a fundamental, provided **that the third interval between** the new tones is also consonant. Now the intervals **within** an **octave** which are consonant are **the** fifth, $\frac{3}{2}$, **the** fourth, $\frac{4}{3}$, the major sixth, $\frac{5}{3}$, the major third, $\frac{5}{4}$, the minor third, $\frac{6}{5}$, and the minor sixth, $\frac{8}{5}$. To these may be added the natural seventh, $\frac{7}{4}$.

Thus the **only consonant** triads or combinations of **three notes** within an octave are :—

(1) $C\ E\ G$.   (2) $C\ E\flat\ G$.
(3) $C\ F\ A$.   (4) $C\ F\ A\flat$.
(5) $C\ E\flat\ A\flat$.   (6) $C\ E\ A$.

From the two first all others **are** derived, and they consist **of two** thirds, one major, one minor, in different order. In the **major** triad, **the** major third **is** below, **the minor** above (1). Whereas in (2) **the** minor third is below, the major above; the chord is a minor triad. Nos. (3) and (4) are termed chords **of** the fourth and sixth. They may **be** considered as having been formed from the two former, by transposing the G an octave **lower.**

The **last** two (5), **(6) are** termed **chords** of the **third and** sixth, or chords **of the** sixth. **They** may be looked **on as** transpositions of **a** major or **minor** chord in which C is placed an **octave** higher. The natural seventh forms no part of **any triad, since,** though **consonant with** C, **it** is dissonant with all other intervals consonant with C.[1]

Relative Harmoniousness.—In adverting **to the** relative harmoniousness of chords, the combinational tones must be remembered, especially those of **the first** order, between the primes and their first upper partials.

---

[1] From this peculiar character, the natural seventh is not used **in** music, and has been designated by X in a former Chapter.

Major triads with their combinational tones :—

Minor triads :—

The primes are marked as minims, their combinational tones as crotchets, those between primes and first upper partials as quavers and semi-quavers. The last in the major triads are practically inaudible in the harmonium, though they may be heard in the louder combinations of the organ.

In the minor triads, the combination tones of the first order are easily audible, and disturb the harmony, as will be seen by reference to the example.

The same law is carried out in consonant triads which exceed the octave, and transposition for the purpose of widening the intervals affects their harmoniousness.

Helmholtz gives as rules :—

1. Those intervals in which the smaller of the two numbers expressing the ratios of the vibrational numbers is even, are improved by having one of their tones transposed an octave, because the numbers expressing the ratio are so diminished, thus :—

The fifth — $2:3$ becomes the twelfth $2:6 = 1:3$.
The major third — $4:5$ becomes the major tenth $4:10 = 2:5$.

2. Those in which the smaller of the two ratio numbers is odd are made worse by the same process. For instance the fourth $3:4$ becomes the eleventh $3:8$, and so on.

By a further extension of this beautiful method, which space does not here permit, it may be shown that the physical theory of consonance and dissonance leads to rules which previous theories could not contain, although they have often been æsthetically and instinctively followed by great composers.

The Scale.—It has been shown that if any note which may be represented by C be played on a musical instrument it introduces, by the harmonic law, two other allied notes, E and G, the vibrations of which stand to those of the first in the simple ratio of 4 : 5 : 6, and form the harmonic triad. Similarly, if G, thus found, be taken as the basis of a triad, it will be followed by B, D_2, bearing the same relation. We should then have the following scale :—

	C	D	E	G	B	C_2
	1	$\frac{9}{8}$	$\frac{5}{4}$	$\frac{3}{2}$	$\frac{15}{8}$	2
	First	Second	Third	Fifth	Seventh	Eighth
Intervals		$\frac{9}{8}$	$\frac{10}{9}$	$\frac{6}{5}$	$\frac{5}{4}$	$\frac{16}{15}$

the fractions representing the ratios of the vibrations of each note to that next below it.

But it is obvious that the space between E and G, as well as the space between G and B, require filling up. D has been suggested as the basis of a new triad, but this note gives very complex results. If, however, C_2 be taken as the upper element of a third triad, the two lower members of which would be F and A, we get an $A = \frac{5}{3}$ and an $F = \frac{4}{3}$, with which we can complete the scale of eight notes with the following intervals and vibration numbers, m representing that of the foundation note :—

	m	$\frac{9}{8}m$	$\frac{5}{4}m$	$\frac{4}{3}m$	$\frac{3}{2}m$	$\frac{5}{3}m$	$\frac{15}{8}m$	$2m$
	C	D	E	F	G	A	B	C
Intervals		$\frac{9}{8}$	$\frac{10}{9}$	$\frac{16}{15}$	$\frac{9}{8}$	$\frac{10}{9}$	$\frac{9}{8}$	$\frac{16}{15}$

Here we have three unequal intervals only employed, which are termed respectively,—

Major tone	$\frac{9}{8}$
Minor tone	$\frac{10}{9}$
Major semitone	$\frac{16}{15}$

This forms a sufficient and satisfactory scale for a single key. But as it is possible to take any other note besides C as the foundation of a scale, terming it the key-note, it becomes necessary to interpolate intermediate sounds between those thus found, so as to preserve the same rotation of intervals. These are five in number, situated between the larger intervals, or whole tones. They are not given new names, but termed the FLATS or SHARPS of the sounds between which they lie. The amount of the flattening or sharpening is in either case represented by the fraction $\frac{25}{24}$, which is termed a chromatic semitone, and which differs somewhat from the major semitone above given.

It may here be noticed that the difference between the major and minor tones is not without importance. This difference may be obtained by inverting one fraction and multiplying it into the other, which is equivalent to division.

$$\tfrac{9}{8} \times \tfrac{9}{10} = \tfrac{81}{80} = \text{COMMA},$$

as this computational interval is termed. Another fractional difference may also be adverted to in this place. The major third from C to E is given above as $= \tfrac{5}{4}$ and the fifth as $\tfrac{3}{2}$. If these be divided into one another the difference $= \tfrac{6}{5}$, which is termed a minor third. Now $\tfrac{5}{4}$ exceeds $\tfrac{6}{5}$ by $\tfrac{25}{24}$.

$$\tfrac{5}{4} \times \tfrac{5}{6} = \tfrac{25}{24} = \text{MINOR SEMITONE}.$$

We thus obtain three semitones of different size. The interpolated flats and sharps may be conveniently tuned to the second or chromatic interval by altering the vibration number in the given proportion. We have thus the usual twelve notes of the scale termed chromatic, constructed on the simple plan that a note is sharpened by increasing its vibrations in the proportion $\frac{25}{24}$, or flattened by diminishing them in the ratio $\frac{24}{25}$. If all the notes of the simple scale be thus treated we obtain twenty-one to the octave which are of sufficient importance to deserve tabulation. The ratios are given in logarithmic form, as this renders the actual steps more obvious than they are in their fractional shape.

SCALES, CHORDS, TEMPERAMENT, Etc.

Table of Twenty-one Notes to the Octave.[1]

	Interval.	Logs. of Ratio.
C	First	·00000
C♯		·02312
D♭		·02803
D	Second	·05115
D♯		·07428
E♭		·07379
E	Third	·09691
E♯		·12003
F♭		·10181
F	Fourth	·12494
F♯		·14806
G♭		·15297
G	Fifth	·17609
G♯		·19922
A♭		·19873
A	Sixth	·22185
A♯		·24497
B♭		·24988
B	Seventh	·27300
B♯		·29612
C♭		·27791
C	Eighth	·30103

It will be observed that D♯ is higher than E♭, E♯ than F♭, G♯ than A♭, and B♯ than C♭, but the notes are all really distinguishable.

Taking in succession all the other naturals as key-notes, we could construct on each a similar scale of twenty-one, or **141** in all, of which many indeed would be identical, but of which about 100 would remain distinct. This number would therefore be needed if every possible key were to be exact.

The various methods of surmounting, or rather compro-

---

[1] Modified from Professor Haughton's *Natural Philosophy*, p. 181.

mising, this difficulty are termed TEMPERAMENTS, and will be considered presently.

Pythagorean Scale.—Pythagoras laid down a somewhat different series of values, according to the following scheme—

C	D	E	F	G	A	B	C
$\frac{9}{8}$	$\frac{9}{8}$	$\frac{256}{243}$	$\frac{9}{8}$	$\frac{9}{8}$	$\frac{9}{8}$	$\frac{256}{243}$	

Here the fourth, fifth, and octave are identical with the ordinary system; the major third, sixth, and seventh are greater by a comma; while the small interval or semitone is diminished by the same quantity. In this system the only numbers that appear are 2 and 3, whereas in the modern, the number 5 appears: hence the interval between any two notes of the Pythagorean scale can be expressed as the sum or difference of a certain number of octaves and fifths. A violin tuned to true fifths really plays the Pythagorean scale, with power, however, of modifying dissonant notes.

Minor Scale.—It was shown that the harmonic triad, consisting of the ratio 4 : 5 : 6, may be broken up into two intervals, denoted respectively by the fractions $\frac{5}{4}$ and $\frac{6}{5}$, which are termed *major* and *minor* thirds. These unequally divide the containing fifth, the ratio of which is the product of these fractions, or $\frac{3}{2}$. In the major scale given above the major third precedes, and is followed by the minor. But if they be transposed, and the minor third taken first, we entirely alter the musical character of the scale, and produce an essentially different sensorial or emotional effect: whereas the major scale has a cheerful and exciting tendency, the minor, to most if not to all hearers, is melancholy and pathetic. The altered position of the component notes, moreover, still further complicates the constituent ratios, and renders the question of temperament even more arduous. The scale thus formed has the ratios as follows:—

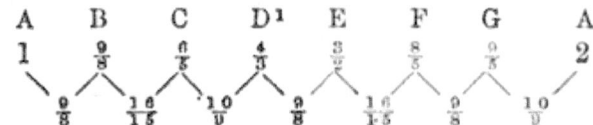

in which a key-note is assumed a minor third below that of the corresponding major scale; A, for instance, instead of C, and so on.

[1] It is to be remarked that the note D the fourth of the above scale, is really a comma too sharp.

Temperament.—It will already have become apparent to the reader that there is an obvious lack of arithmetical agreement between the various intervals as represented fractionally. The cause of this lies deep in the nature of numbers, and is well expressed by Mr. Ellis. "It is impossible," he says, in an appendix to his translation of Helmholtz's work, "to form Octaves by just Fifths or just Thirds, or both combined, or to form just Thirds by just Fifths, because it is impossible by multiplying any one of the numbers $\frac{3}{2}$ or $\frac{5}{4}$ or 2, each by itself, or one by the other, any number of times, to produce the same result as by multiplying any other one of these numbers by itself any number of times." The physical fact may be otherwise stated by saying that the octave and the fifth are incommensurable, just as are the diameter and circumference of the circle.

The simplest way of representing this incommensurability is to take a case. If the octave be divided into twelve equal semitones, the fifth ought to be seven of these; but it was found out very early in the history of music that a fifth is a little more than seven. It is about 7·01955. Consequently, taking twelve of these fifths, they give rather more than seven octaves. They do not return to the corresponding octave of the starting note. The difference or departure is the above figure multiplied by 12, or 0·23460 of a semitone. This old discovery is generally attributed to Pythagoras, and the figure 0·23460 is termed the "Comma" of Pythagoras. Whether Pythagoras deserves the credit of the discovery, or whether he imported it from Egypt, is matter of doubt; but, at any rate, the Greeks knew of the monochord, of the ratios to be derived from it, and of the divisions of the scale. Euclid wrote a work called the *Sectio Canonis*, or the Division of the String, which enters into full details. The third of the Greek scale was made by four fifths taken upwards, and is still called a Pythagorean third. In the same way six major tones exceed an octave by the Pythagorean comma.

It should be distinctly noticed that this discrepancy is a law of nature, not inherent in any particular system or method, and entirely beyond man's control.

Temperament may be defined as *the division of the octave into a number of intervals such that the notes which separate them may be suitable in number and arrangement for the purposes of practical harmony.* The possibility of any other division than that recognised in the ordinary piano and harmonium will be new to many readers; for the usual form of keyed instrument is so engraven on our mind that most persons are unaware that any other arrangement exists. The common instrument has of course its own system of temperament, one that, though not the oldest, is certainly the simplest, and which is usually termed equal.

Equal temperament aims at dividing the octave into twelve equal parts or semitones. If it so happened that the octave could be divided thus, and the other intervals, such as the fifth and third, retained in tune, it would be a great boon. Unfortunately nature has not so ordained it.

The attempts to remedy this inherent incommensurability of the musical scale have been numerous and varied, some dating back to ancient times, others of very modern construction. Their principal varieties may be given best in a tabular form. They have chiefly been applied to keyed instruments such as the organ, piano, and harmonium, where their necessity is most felt.

Table of Temperaments.

1. Systems retaining the ordinary keyboard.
    a Unequal or mean-tone temperament.
    b Equal temperament.

2. The ordinary keyboard with additional keys.
    a Handel's Foundling organ.
    b The old Temple organ.
    c The digitals of the concertina.

3. Additional keyboards.
    Helmholtz's harmonium with Gueroult's modifications.

4. Additional keys and keyboard.
    Perronet Thompson's enharmonic organ.

5. The ordinary keyboard with combination stops.
    Mr. Alexander Ellis's harmonium.

6. Entirely new arrangement of keyboard.
    a Poole's system.
    b Bosanquet's generalised keyboard.
    c Colin Brown's voice harmonium.

1. (*a*) The old unequal or MEAN-TONE SYSTEM was an attempt to get the more common scales fairly accurate,

leaving those less needed out of account, the most faulty being termed "wolves." There was a consequent condition in dealing with this tuning that the player should limit himself to a prescribed circle, and should never modulate into forbidden keys. The temperament had many merits, and some organists even now prefer it to the equitonic system. Its principle is as follows.[1] If we take four exact fifths upwards they lead to a third a comma sharper than the perfect third. If then we make each of the four fifths one-fourth of a comma flat, the resulting third is depressed a whole comma, and coincides with the perfect third. This is the rule of the mean-tone system. It is so called because its tone is the arithmetic mean between the major and minor tones of the diatonic scale. It can be traced back to two Italian authors of the sixteenth century, Zarlino and Salinas, from which time it spread slowly, and about 1700 was in universal use. It was employed by Handel and his contemporaries, and kept its ground in this country until within the last few years. Many organs were till lately tuned on this plan, and some, such as that in St. George's Chapel, Windsor, still remain. The change from this to the equal temperament is generally supposed to be due to the influence of Bach, though Mr. Bosanquet, in the work above quoted, adduces some strong evidence to the contrary. The differences of the old and equal systems, and their respective departures from just intonation, may be seen in a compact form from an abbreviation of Ellis's table as follows:—

Note.	Old.	Just.	Equal.
C	30,103	30,103	30,103
B	27,165	27,300	27,594
A	22,320	22,185	22,577
G	17,474	17,609	17,560
F	12,629	12,494	12,545
E	9,691	9,691	10,034
D	4,846	5,115	5,017
C	0	0	0

[1] *An Elementary Treatise on Musical Intervals and Temperament*, by R. H. M. Bosanquet.

(*b*) According to the EQUAL TEMPERAMENT the octave is divided into twelve perfectly similar intervals, seven of which are taken for the fifth, although its real measure is $7\frac{1}{11}$ of these. It is thus somewhat flattened, from 17,609 to 17,560, by the interval termed a *Schisma* = 49, though less so than in the older system, which lowers it to 17,474, as will be seen from the table. On the other hand the third is far too sharp, 10,034, or nearly two-thirds of a comma, instead of 9,691 as it stands correctly in both the other columns. The sixth, moreover, is rendered extremely sharp in equal temperament, namely 22,577, or eight schismas, as against the true 22,185. The seventh is flattened in the old more than two schismas, and considerably sharpened in the equal method, by exactly six schismas. The fourth of the scale is less altered in proportion to its sensitiveness, being raised rather more than two schismas in the old, and only one in the equal system. The second of the scale stands in a peculiar position, being a double note. The old temperament places it about halfway between its grave and acute forms, whereas the equal method removes it nine schismas above the grave form, thus constituting the largest departure from accuracy to be met with.

(2, *a*) Even as early as the time of Handel the advantage to be derived from additional keys was obviously appreciated, for it is known that he presented to the Foundling Chapel an organ thus furnished. (*b*) The original organ in the Temple Church, built by Father Smith in 1688, possessed fourteen sounds to the octave instead of twelve, the A♭ and G♯ as well as the E♭ and D♯ being distinct and divided.[1] The keys themselves were split across in the middle, the back halves rising above the front portions, so that the finger could be placed on either at the player's discretion. The range of good keys on the unequal system was thus materially extended.[2] (4) The device of additional keys was, however, carried to its fullest development by Colonel Perronet Thompson in his enharmonic organ, which may still be seen at the South Kensington Museum. He used the large number of seventy-two to the octave, which were further distributed on three different keyboards, but which also differed among themselves in colour, shape, appearance, and in name. Besides

---

[1] See Curwen's *Tract on Musical Statics*, pp. 11, 103.
[2] (*c*) The same contrivance has been applied to the just English concertina, which is tuned to the mean-tone temperament, with duplicate studs for D♯—E♭ and G♯—A♭.

VII.]   SCALES, CHORDS, TEMPERAMENT, Etc.   147

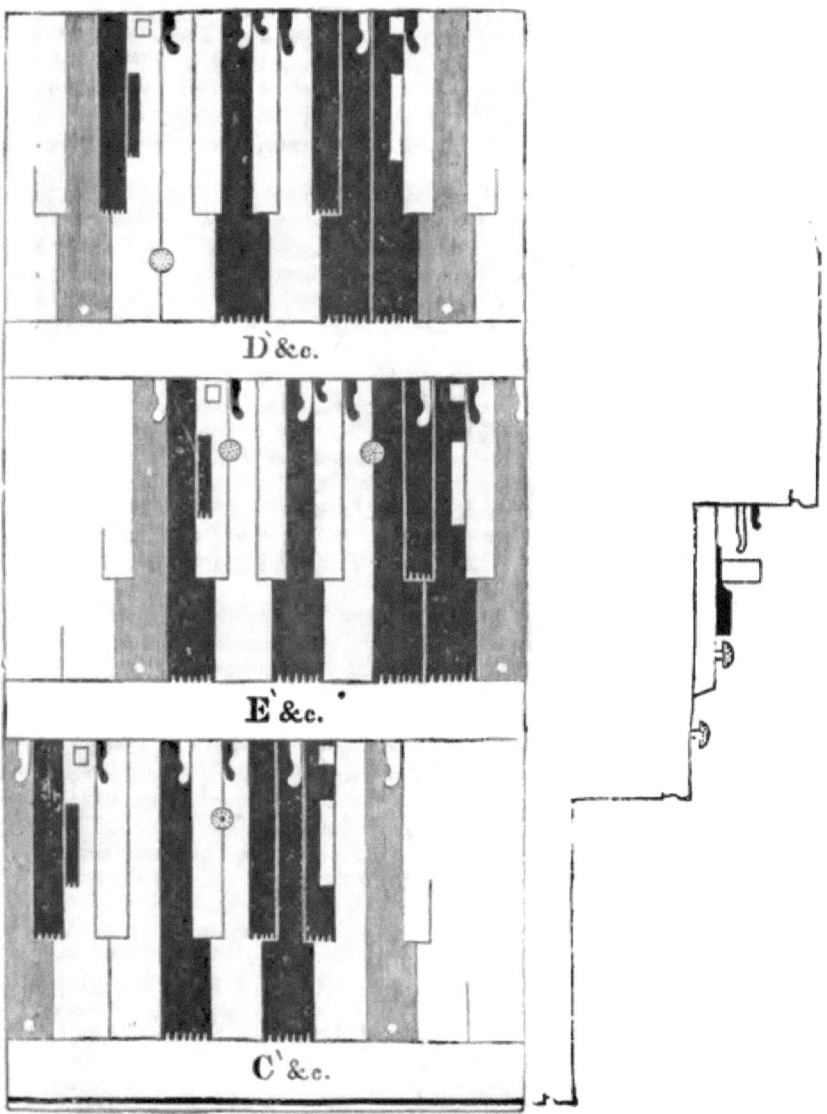

Fig. 52.—Perronet Thompson's Keyboard.

L 2

the ordinary digitals there were others termed *Flutals*, *Quarrills*, and *Buttons*. By this means, though still retaining the ordinary arrangement of the keyboard, he was enabled to produce accurately twenty-one scales with a minor to each of them. He employs a cycle of fifty-three sounds, of which he uses about forty, the full cycle being discontinued at a certain point.

(3) The difficulty of adding new sounds without undue mechanical complication has been attacked in a different way by Helmholtz. The keyboards are in this case increased to two, so as to obtain twenty-four instead of twelve notes to the octave. They are of half the usual depth, placed one above the other, as in the organ. This has always seemed to the writer a practical and simple system. The instrument made for Helmholtz was so tuned that all the major chords from $F\flat$ to $F\sharp$ could be played on it. On the lower manual were the scales from $C\flat$ major to G, and on the upper those from $E\flat$ major to B major. To modulate beyond B major on one side and $C\flat$ major on the other it was necessary to make the enharmonic change between these two notes, which perceptibly alters the pitch by the interval of a comma, $\frac{81}{80}$. The minor modes on the lower manual were B or $C\flat$ minor, on the upper $D\sharp$ or $E\flat$ minor.

(4) The same idea has been carried out with slight variation in an instrument shown at South Kensington, namely Gueroult's modification of Helmholtz's harmonium, of which the following is the maker's own description.

This instrument has a front and back keyboard, each divided into twelve semitones, like that of a piano, and each possessing five octaves. They are both tuned to true fifths, but the back keyboard is throughout a comma flatter than the front, which is on the normal diapason. The black keys on each keyboard therefore do duty for a flat and a sharp, but not in the same series. On the front keyboard, for instance, $E\flat$ represents the $D\sharp$ of the back. Considered as flats, the black keys of the second keyboard represent sharps of a third board which would be tuned a comma lower than the second. By thus fusing the flat of one series with the sharp of the other, an error is committed equal to the interval $\frac{846}{845}$,[1] which is at the extreme limit of audible differences.

On the front keyboard, starting from A, the following notes are tuned to true fifths, so as to give no beat whatever: A, E; E, B; D, A; G, D; C, G; F, C; $B\flat$, F; $E\flat$, $B\flat$. The

---

[1] In Helmholtz, $\frac{886}{885}$.

perfect chords D, F♯, A ; A, C♯, E ; E, G♯, B are also made. The fifths D, A ; A, E ; E, B are those previously determined ; the F♯, C♯, and G♯ are the thirds which give no beat in the perfect chords.

On the back keyboard B, E, A, D, G, F, B♭ are in succession fixed by taking these notes as true thirds in the perfect chords G, B, D ; C, E, G ; F, A, C ; B♭, D, F ; E♭, G, B♭ ; D♭, F, A♭ ; G♭, B♭, D♭ ; of which the fifths C, G, F, &c., are taken on the front keyboard. The chords D, F♯, A ; A, C♯, E ; E, G♯, B, are formed on the back keyboard, using for the fifths D, A, E, B sounds already found, and tuning the thirds without beats.

The B of the back keyboard forms a true fifth with the F♯ or G♭ of the front. The D♯ or E♭ of the back keyboard is got by taking it as the true third of the perfect chord B, D♯, F♯, the two first notes being taken on the back and the third on the front board. The G♯ or A♭ of the front board gives, with the E♭ of the same board, a fifth which is not quite true, being exactly equal to the tempered fifth. The C of the back board is determined by so taking it that the resultant tones of the two thirds about it should be free from beats. In the six major scales of C, F, B♭, E♭, A♭, D♭ the fingering is the same, the third, sixth, and seventh are on the back keyboard, all the others on the front. The keys of G, D, A are played with the sharpened notes on the front board. The key of G has thus only B and E on the back board, of D only B, and A has none. A can be played entirely on the back board also. The key-notes of all minor scales are on the back board. For the minor scales of A, D, G, C, F, and B♭ the third and sixth alone are on the front keyboard.

(5) A somewhat simpler method of working Helmholtz's system has been suggested by Mr. Alexander Ellis, and carried out by Mr. Saunders. The keyboard is single, but communicates with two rows of vibrators tuned according to the method given above, or in the following series :—

Back row	B♯	D♭	C♯♯	E♭	F♭	E♯	G	F♯♯	A♭	G♯♯	B♭	C♭
Front row	C	C♯	D	D♯	E	F	F♯	G	G♯	A	A♯	B

When no stops are drawn out, the arrangement is that of the front series, the white notes being naturals and the black sharps. On pulling out a stop, the vibrators of its name in the front series of the instrument are damped, and the corresponding vibrators of the back series come into action, until the notes speaking are those of the old-fashioned manual. Between these extremes any required combination of notes can be produced, from seven flats to seven sharps, according to the keys employed. This method, which entirely removes the difficulties of complex fingering, has the disadvantage of requiring a constant alteration of stops, which in transitory modulations is occasionally laborious.

(6) The last class of contrivance for producing true intonation does away with the ordinary form of keyboard altogether. It is impossible here to give full details of these instruments, which practically introduce a new principle into musical execution. Poole's, Bosanquet's, and Colin Brown's forms may be taken as typical representatives of many less perfect devices. In all, the series of tones are arranged diagonally one beyond another, so "that the form of a chord of given key relation is the same in every key. But the notes are not all symmetrical, and the same chord may be struck in different forms according to the view which is taken of its key relationship."[1] They therefore possess the great advantage of similarity of manipulation, although this is quite different from that ordinarily taught. It would appear, however, that the new systems are far from difficult to learn by any person who has obtained some experience on the older form of instrument.

(*a*) The first attempt in this direction was made by H. W. Poole, of South Danvers, Massachusetts, U.S. The instrument appears to have been constructed, and is described in *Silliman's Journal* for 1850. His organ was intended to contain 100 pipes to the octave, and the scale to consist of just fifths and thirds in the major chords, and also the natural or harmonic sevenths. The arrangement of keys is best given by a diagram extracted from Mr. Bosanquet's work.

[1] Bosanquet, *op. cit.* p. 45.

VII.] SCALES, CHORDS, TEMPERAMENT, Etc. 151

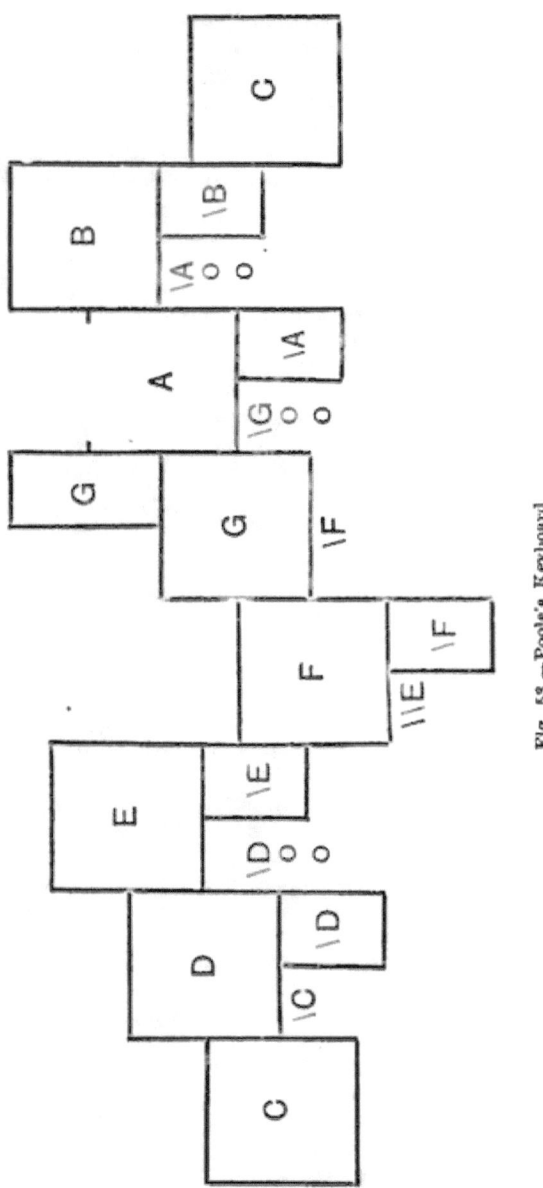

Fig. 53.—Poole's Keyboard.

According to Mr. Bosanquet's notation here used, notes are arranged in series in order of successive fifths. Each series contains twelve fifths from F♯ up to B. One series is *unmarked*. It contains the standard C. Each note of the next series of twelve fifths up is affected with the mark /, which is called a mark of elevation, and is drawn upwards in the direction of writing. The next series has the mark //, and so on. The series below the unmarked series is affected with the mark \, which is called a mark of depression, and is drawn downwards, in the direction of writing; the succeeding series is marked \\, and so on. Where, as in Poole's keyboard, perfect thirds are tuned independently of the fifths, they are here represented by the note eight fifths distant in the series; this is a close approximation to the perfect third, according to a relation which has been called Helmholtz's Theorem. Thus C—\E means a perfect Third; \E—\G♯ is also a perfect third (chord of dominant of \A minor). The places of harmonic sevenths are marked by circles (O).[1]

(b) *Bosanquet's Generalised Keyboard.*

In the enharmonic harmonium exhibited at the Loan Collection of Scientific Instruments, South Kensington, 1876, there was a keyboard which can be employed with all systems of tuning reducible to successions of uniform fifths; from this property it has been called the generalised keyboard. It will be convenient to consider it first with reference to perfect fifths. These are actually applied in the instrument in question to the division of the octave into fifty-three equal intervals, the fifths of which system differ from perfect fifths by less than the thousandth part of an equal temperament semitone.

It will be remembered that the equal temperament semitone is the twelfth part of an octave. The letters E. T. are used as an abbreviation for the words "equal temperament."

The arrangement of the keyboard is based upon E. T. positions taken from left to right, and deviations or departures from those positions taken up and down. Thus the notes nearly on any level are near in pitch to the notes of an E. T. series; notes higher up are higher in pitch; notes lower down lower in pitch.

The octave is divided left to right into the twelve E. T. divisions, in the same way, and with the same colours, as if the broad fronts of the keys of an ordinary keyboard were removed, and the backs left.

[1] *Proceedings of the Musical Association*, 1874-75, p. 14.

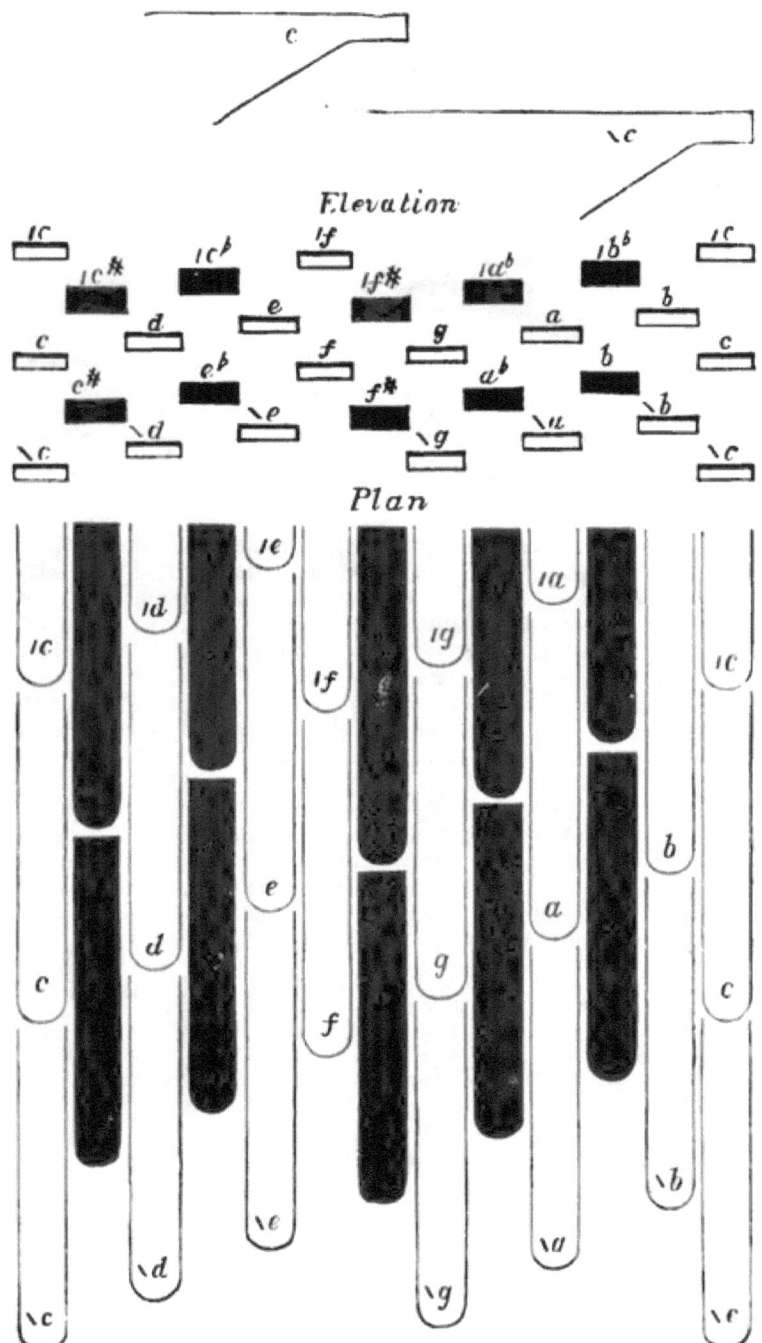

Fig. 54.—Bosanquet's Generalised Keyboard.

The deviations from the same level follow the series of fifths in their steps of increase. Thus G is placed one-fourth of an inch further back, and one-twelfth of an inch higher, than C; D twice as much, A three times, and so on, till we come to /C, the note to which we return after twelve fifths up; this note is placed three inches further back, and one inch higher, than the C from which we started.

With the system of perfect fifths the interval C—/C is a Pythagorean comma. With the same system, the third determined by two notes eight steps apart in the series of fifths (C—\E) is an approximately perfect third. With the system of fifty-three the state of things is very nearly the same as with the system of perfect fifths.

The principal practical simplification which exists in this keyboard arises from its arrangement being strictly according to intervals. From this it follows that the position-relation of any two notes forming a given interval is always exactly the same; it does not matter what the key relationship is, or what the names of the notes are. Consequently a chord of given arrangement has always the same form under the finger; and, as particular cases, scale passages as well as chords have the same form to the hand in whatever key they are played, a simplification which gives the beginner one thing to learn, whereas there are twelve on the ordinary keyboard.

The keyboard has been explained above with reference to the system of perfect fifths and allied systems; but there is another class of systems to which it has special applicability, the mean-tone and its kindred systems. In these the third, made by tuning four fifths up, is perfect or approximately perfect. The mean-tone system is the old unequal temperament. The defects of that arrangement are got rid of by the new keyboard, and the fingering is remarkably easy. The unmarked naturals in the diagram present the scale of C when the mean-tone system is placed on the keys.[1]

*(c) Colin Brown's **Natural Fingerboard** with **Perfect Intonation**.*

The digitals consist of three separate sets, of which those belonging to four related keys, representing the notes 2, 5, 1, 4, are white; those belonging to three related keys, and

[1] For further details see *An Elementary Treatise on Musical Intervals and Temperament, with an account of an Enharmonic Harmonium exhibited in the Loan Collection of Scientific Instruments, South Kensington, 1876; also of an Enharmonic Organ exhibited to the Musical Association of London, May, 1875, by R. H. M. Bosanquet, Fellow of St. John's College, Oxford.* (London: Macmillan, 1876.)

representing 7, 3, 6, are coloured; the small round digitals represent 7 *minor*, or the major seventh **of the minor scale.** These are the same in all keys.

This fingerboard can be made to consist of any number of keys. The **scales** run in the usual order in direct line horizontally from left to right *along* the fingerboard.

The keys **are at** right angles to the scales, and run vertically *across* the keyboard, from \C♭ in the front to /C♯ at the back, C being the central key.

The scale to be played is always found in direct line horizontally between the key-notes marked on the fingerboard, but the digitals may be touched at any point.

The order of succession is always the same, and consequently the progression of fingering the scale is identical in every key.

The first, second, fourth, and fifth tones of the scale are played by the white digitals, the third, sixth, and seventh by the coloured.

The sharpened sixth and seventh of the modern minor scale are played by the round digitals. The round digital, two removes to the left as in the key of B flat, is related to that in the key of C as 8 : 9, and supplies the sharpened sixth in the relative minor of C; so in all keys similarly related.

Playing the scale in each key the following relations appear (see diagram on p. 157):—

From white digital to white, say from the first to second and fourth to fifth of the scale, and from coloured to coloured, or from the sixth to the seventh of the scale, the relation is always 8 : 9.

From white to coloured, being from the second to the third, and from the fifth to the sixth of the scale, 9 : 10.

From coloured to white, being from the third to the fourth, and from the seventh to the eighth of the scale, 15 : 16.

From *white* to *white*, or *coloured* to *coloured*, is always the *major tone*, 8 : **9.**

From *white* to *coloured* is always the *minor tone*, 9 : 10.

From *coloured* to *white*, the diatonic semitone, 15 : 16.

The round digital is related to the coloured which succeeds it as 15 : 16, and to the white which precedes it as 25 : 24, being the imperfect chromatic semitone.

Looking *across* the fingerboard at the digitals *endwise*, from the end of each white digital to the end of each coloured immediately above it, in direct line, the relation is always 128 : 135, or the chromatic semitone; and from the end of each coloured digital to the white immediately above it, in direct line, the comma is found, 80 : 81.

Between all enharmonic changes, such as between A flat 404♯♯ to G sharp 405, the interval of the schisma always occurs, 32,768 : 32,805, the difference being 37.

These simple intervals and differences, 8 : 9, 9 : 10, 15 : 16, 24 : 25, 80 : 81, 128 : 135, and 32,768 : 32,805, comprise all the mathematical and musical relations of the scale. The larger intervals of the scale are composed of so many of 8 : 9, 9 : 10 and 15 : 16, added together. The "comma of Pythagoras," being a comma and schisma added together, is found between every enharmonic change of key, as from \C♭ to /B, or twelve removes of key.

The digitals rise to higher levels at each end, differing by chroma and comma, or comma and chroma, alternately. This causes separate levels on the fingerboard at each change of colour. Though these are not essential, they will be found very useful in manipulation, and serve readily to distinguish the different keys.

The two long digitals in each key are touched with great convenience by the thumb. The lower end of each coloured digital always represents the seventh in its own key, and the borrowed, or chromatic sharp tone, in every other; thus the seventh in the key of G is the sharpened fourth, or F sharp, in the key of C; and so in relation to every other chromatic sharp tone.

The white digital is to every coloured digital as its chromatic flat tone; thus the fourth in the key of F is B flat, or the flat seventh in the key of C; so in relation to every other chromatic flat tone. In this way all chromatic sharp and flat tones are perfectly and conveniently supplied without encumbering the fingerboard with any extra digitals, such as the black digitals on the ordinary keyboard, the scale in each key borrowing from those related to it every possible chromatic tone in its own place, in perfect intonation. The tuning is remarkably easy, and as simple as it is perfect.

It will be observed that each line of digital across the fingerboard bears one generic name, as C♭, C, C, C♯, C♯. So with every other line.

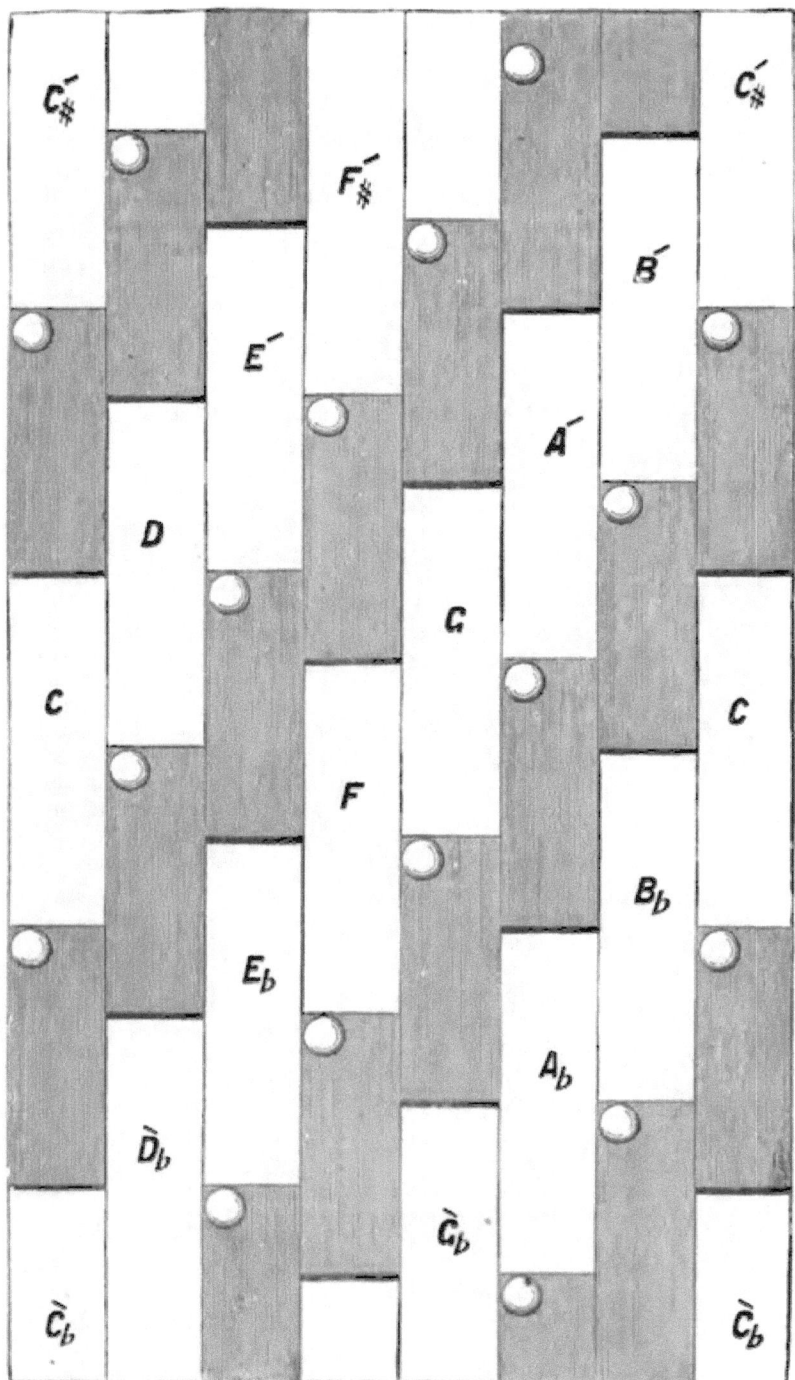

Fig. 5b.—Plan of Natural Fingerboard.

While all the major keys upon the fingerboard, according to its range, have relative minors, the following, |\ B♭, \F, \C, \G, \D, A, E, B, F♯, C♯, G♯, and D♯, can all be played both as major and as perfect tonic minors.

These secondary keys are more than appear at the first inspection of the fingerboard. A series of round digitals placed upon the white, and a comma higher, additional to those placed upon the coloured digitals, would supply the scale in every form the most exacting musician could desire, but it is a question if such extreme extensions are either necessary or in true key relationship, and whether simplicity in the fingerboard is not more to be desired than any multiplication of keys which involve complexity and confusion.[1]

Adaptation of True Intonation to the Orchestra.—There can be no doubt that the place in which the superiority of true over tempered intonation is most felt, and in which it could be most easily attained, is the orchestra. Unfortunately this is exactly where it has been most thoroughly neglected. To a certain degree it is instinctively and unconsciously produced; the stringed instruments having the power of modifying their notes accurately by ear according to the requirements of a particular chord or modulation. Even here, however, the result is marred by the erroneous practice of many violinists. But in the wind-instrument department this power is limited to the slide trombones, and the corresponding trumpet. It is much to be regretted that the natural and perfect quartett formed by the slide trumpet, alto, tenor, and G bass trombones should be so much disfigured at the present day, the alto being almost always replaced by a second tenor, the bass often omitted or transferred to the euphonium, and the inimitable trumpet spoiled by the cornet. But beyond these two types no endeavour whatever seems to have been made towards true intonation. Flutes, oboes, clarinets, and bassoons, as made and supplied to their respective players, are rarely in tune at all, even to themselves, and, at most, present a feeble approach to equal temperament. There is no insuperable obstacle in the way of their all being so manufactured and tuned as to give at least the principal enharmonic differences, by utilising duplicate fingerings for the same note, which already exist on all of them.

---

[1] A full description of the voice harmonium may be found in the specification for patent. The principles upon which it is constructed and tuned will be found fully stated in *Music in Common Things* parts i. and ii., published by Messrs. William Collins, Sons, and Co., Glasgow, and Bridewell Place, New Bridge Street, London, and the Tonic Solfa Agency, 8, Warwick Lane, E.C.

The writer has himself directed the construction of a clarinet and bassoon which, without any appreciable increase of mechanism, and without the slightest alteration of the system of fingering usually adopted, can be made to give nineteen notes to the octave, a number more than sufficient to provide for all the commatic differences involved. The clarinet, three forms of which are commonly used in the orchestra, lends itself particularly well to such an arrangement, from the fact that it is rarely if ever required to play in high flat or sharp keys. Those actually in use can therefore be tuned accurately, and a duplicate note provided in case of sudden and temporary modulations.

It is, however, essential that the present music should be carefully gone over in score by a competent harmonist, and the modified notes marked into the band parts according to Mr. Bosanquet's or any equally good notation, with a mark of elevation or depression, according to their specific key relationship. An orchestra in which perfect intonation were thus secured would instantly obtain what is very obvious in listening to keyboard instruments of correct scale, such as that of Mr. Colin Brown, namely, a largely increased volume of tone in proportion to the instrumental resources employed.

An excellent adaptation of the enharmonic principle to brass valved instruments has been devised by Mr. H. Bassett, F.C.S., in what he terms the COMMA and TELEOPHONIC trumpets. In all instruments furnished with the ordinary valves there are great faults in intonation. "It is not difficult," says Mr. Bassett, "to show, by calculation from the varying lengths of tube brought into action by the valves, that many of the intervals resulting from their combination are not in accordance with the just or tempered scales. The unfortunate practice of transposing parts written in widely different keys, so as to use only one or two crooks, greatly increases these errors, besides sacrificing the benefit of the natural intervals, and distinctive quality of tone of the different crooks."[1]

He first constructed a valve trumpet in which these faults of intonation should be avoided. In it the first and second valves remain as usual; that is to say they *lower* the pitch by the intervals of a major tone and a diatonic semitone respectively. The third valve *raises* the pitch of any note produced on the first valve by the interval of a comma, or, in other words, the first and third valves *together* lower the pitch a minor tone. This system of valves, which is also applicable

[1] *Proceedings of the Musical Association*, 1876-77, p. 140.

to the French horn, enables the player to produce a practically perfect diatonic scale in the tonic, dominant, and subdominant keys, with the advantage of having only two valve slides to tune when changing the crook, the alteration theoretically required in the third or "comma valve" being so small as to be inappreciable.

In his second, which he terms the "teleophonic" instrument, he retains the original slide, thus keeping the power of adjusting each note to accurate intonation; but he adds a single valve tuned in unison with the open D, or harmonic ninth—in other words, lowering the pitch a minor tone. This valve is worked by the forefinger of the left hand, the instrument being held exactly in the usual manner. By the use of this single valve and the slide, separately or together, it is possible to produce a complete scale, major or minor, with a perfection of intonation limited only by the skill of the player. The valve not only supplies those notes which are false or entirely wanting on the ordinary slide trumpet, including the low A flat and E flat on the higher crooks, but it greatly facilitates transposition and rapid passages.

CHAPTER VIII.

SPECIAL APPLICATIONS TO MUSIC—THE EAR AND VOICE.

Special Applications to Music.—Hitherto Sound has been considered principally in its physical aspect, with only casual reference to its musical application. But a work like the present, though precluded from entering into much detail, would be incomplete without some account of the appliances, vocal and instrumental, out of which the ancient art of music has been constructed. Even in the strictest sense, these may be regarded as apparatus; and their fabrication has in many instances preceded and cleared the way for scientific examination of their mode of operation.

Helmholtz, in drawing distinctions between the physical and æsthetical branches of acoustics, begins by pointing out as a fundamental proposition that "The system of scales, modes, and harmonic tissues, does not rest solely upon unalterable natural laws, but is at least partly also the result of æsthetical principles, which have already changed, and will still further change, with the progressive development of humanity." This, however, does not prevent their being brought under some general law. He divides all musical history into three periods:—

1. The homophonic, or unison music of the ancients, still retained by Orientals and Asiatics.
2. The polyphonic music of the Middle Ages.
3. Harmonic, or modern music, dating from the sixteenth century.

The first, in any extended form, he shows to be only possible in connexion with poetry, such as the hymns and tragedies of Greece. Even in ordinary conversation, however, the voice goes through certain cadences, which are a form of continuous melodic recitation, bearing specific significance in particular forms and phrases. Familiar examples are the falling inflection of the voice at the end of a phrase, and the rising cadence of interrogation.

Polyphonic music appeared in the form of *discant*, in which different voices, each proceeding independently and singing its own melody, had to be united in such a way as to produce either no dissonance, or only transient dissonances, which were readily resolved. In this way canonic imitation arose, as early as the twelfth century. The old ecclesiastical modes were retained, and in 1547 Glareanus distinguished twelve of them, six authentic, and six plagal, assigning to them, somewhat incorrectly, the Greek names, Ionic, Doric, Phrygian, Lydian, Mixolydian, and Æolic, by which they have been known since.

Harmonic or modern music is marked by the independence of its construction, and the artistic connection of its parts; with this system it has become possible to compose works of greater extent, and more energetic expression than its predecessors. Its essential law is that "the whole mass of tones, and the connection of harmonics must stand in a close and perceptible relationship to some arbitrarily selected tonic; and that the mass of tone which forms the whole composition must be developed from this tonic, and must finally return to it."

Notation of Musical Tones.—Several systems of notation have been proposed with a view to distinguish the different octaves from one another. That adopted by Helmholtz, and commonly used in Germany, begins with the great or 8-foot octave, the C of which would be given by a pipe of this size, or by the lowest string of the violoncello. This is marked by capital letters as follows:—

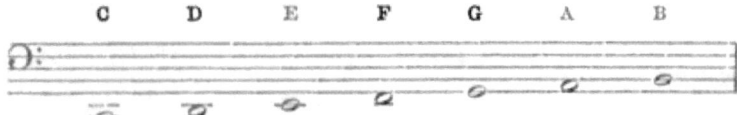

The next octave above this is termed the small or unaccented octave of 4 feet pitch distinguished by small letters.

The 2-foot octave has small letters marked with a single accent thus, c' d' e' f' g' a' b' and is termed the once-accented octave. The 1-foot octave has two of these accents c'' and is termed the twice-accented octave.

Below the great octave is the 16-foot or contra-octave, distinguished by capital letters with an inverted accent below them, thus $C_{,}$ $D_{,}$ $E_{,}$ $F_{,}$ $G_{,}$ $A_{,}$ $B_{,}$ and **below** this again the still deeper, or 32-foot octave, which is marked by capitals with two inverted accents $C_{,,}$ $D_{,,}$ $E_{,,}$ $F_{,,}$ $G_{,,}$ $A_{,,}$ $B_{,,}$.

Mr. Ellis suggests another and very convenient notation **founded on** the powers of 2. "All musicians," he says, "are familiar with the octave, and accustomed to divide the whole range of musical sounds thus. This amounts to selecting a series of tones on the principle of continually multiplying the corresponding number of vibrations by 2. Arithmetically we **are** therefore bound to begin with less than 2 vibrations in a second, which, not being multiplied by two at all, may be said to commence the *zero octave*, and the simplest such number that **can** be selected is 1 itself. Then from 2 up to 4 we multiply the former number **of** vibrations once by 2 and have the first octave. At 4 or twice 2 we multiply twice by 2, and **have** the second octave. The octaves will correspond with **the** number of vibrations with which they begin, thus :—

Oct. 0, 1, 2, 3, 4, 5, 6, 7, 8, 9, 10, 11, 12.
Vib. 1, 2, 4, 8, 16, 32, 64, 128, 256, 512, 1024, 2048, 4096.

The number of the octave is the index of the power of 2, giving the **number** of vibrations with which the octave begins. Thus $256 = 2^8$ **begins** the 8th octave.

Supposing the **notes** C, D, E to occur in any octave, the number of the octave is prefixed. Thus $8E$ means E in the 8th octave. The difficulties arising from using different standards of pitch and temperaments are met by using some fraction as $1\frac{1}{32}$ in place of 1 for the initial number of vibrations. **But** what is commonly called the theoretical pitch, $8C = 256$ vib., is the **only one** suggested by pure arithmetic.

This simple contrivance **for** marking the octave obviates a vast **num**ber of difficulties.

$4C = 16$ vib. **means,** C in the sub-contra octave, or double under-accented **or** underlined great octave $C_{,,}$ or $\underline{C}$, or twice indexed **great** octave C_{2}, or twice negatively indexed **great octave** C_{-2}, or C^{-2}, **or** 32-foot octave, or CCC octave.

$5C = 32$ vib. means C in the contra octave, or an octave above all those named, or CC.

$6C = 64$ vib. means C in the 8-foot octave, the lowest note of the violoncello.

$7C = 128$ vib. means 4-foot C or small c, the lowest note of the viola or tenor violin.

$8C = 256$ vib. means C in the once accented or 2-foot octave, or c', or $\overline{c}$ overlined, or cc with two small c's commonly called "middle C," being on the ledger line between the bass and treble staves.

Varieties of Organ pipes.—Pipes are employed in the organ which aptly illustrate many of the principles previously mentioned. As regards material, they are either of metal or of wood; the former being composed of tin, with a greater or less admixture of lead, sometimes when very large they are made of zinc; the latter of pine wood, cedar or mahogany. In both cases they are divided into open and stopped, as before described, with an intermediate form termed half-stopped; the stopped pipe speaking an octave lower than an open pipe of the same length. But the most important classification is into flue or flute pipes, reeds, and mixtures. The two first have already been fully adverted to; the mixture stops, instead of a single pipe sounding to each note, possess several, from two or three, up to a much larger number, such as twelve or even fourteen. These are arranged in tiers upon the same supply of wind, and thus all sound together. They are tuned to the individual tones of the harmonic series given above, or to some of their octaves, such as the third, fifth, and eighth of the foundation-tone. The sesquialtera, one of the most usual mixtures, was originally a stop of two ranks only, composed of the twelfth and seventeenth intervals above the ground-tone, thus giving prominence to the third and fourth upper partials. The sesquialtera is now often made with three up to six ranks of open metal pipes. The mixture proper is more shrill and acute, comprising the seventeenth, nineteenth, twenty-second, twenty-sixth, and twenty-ninth. As in the treble, the pipes become very small, keen, and prominent, the smaller ranks are discontinued about middle C, and larger pipes, sounding an octave lower, are substituted. This alteration is called a *break*, and takes place also in the sesquialtera. The object of such stops is obviously to increase the brilliancy of effect by reinforcing the upper partial tones which Helmholtz has of late years shown to be always present in sound of a melodious quantity. It is very remarkable that the fact itself had been long ago discovered as a

VIII.]   SPECIAL APPLICATIONS TO MUSIC.   165

matter of Art and Æsthetics, though the scientific solution of the problem on which it depended is modern. Old organs

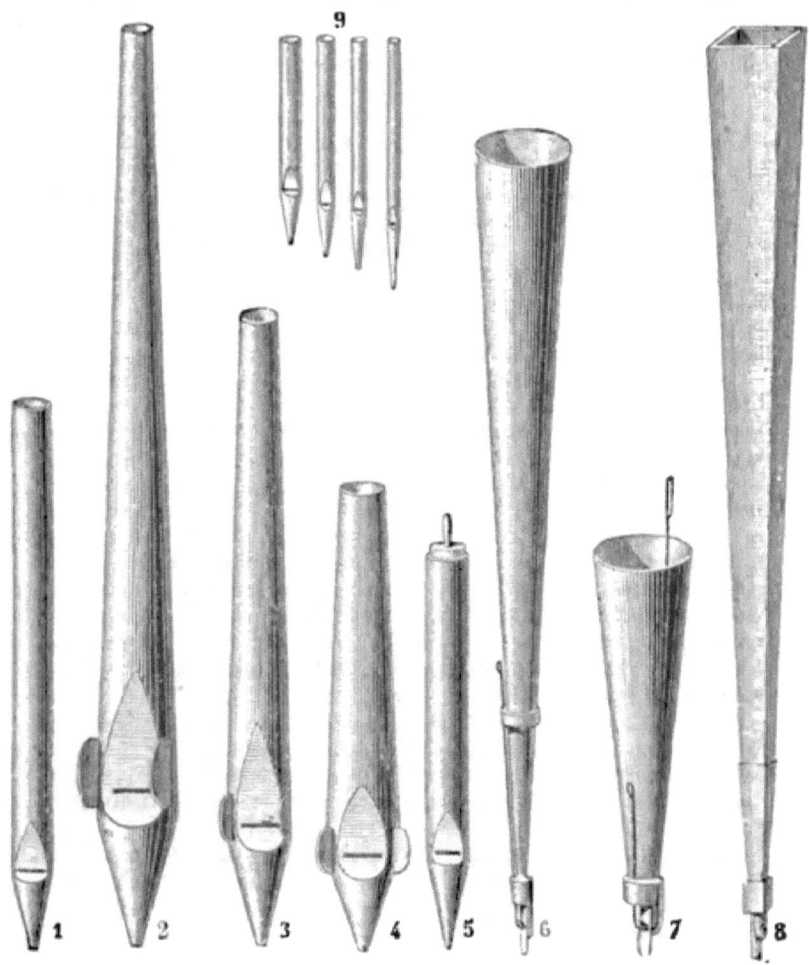

Fig. 56.—Organ stops.

1. Principal (4 feet).—2. Spitz-flöte (8 and 4 feet).—3. Twelfth (3 feet).—4. Cornet.—5. Flute (8 and 4 feet).—6. Trumpet (8 and 4 feet).—7. Vox humana (8 feet).—8. Bombarde, or double reed (16 and 8 feet).—9. Mixture (4 ranks).

possess a mixture stop termed the "cornet" or "mounted cornet" (it being planted or "mounted" on a small sound-board of its own), in which the principle of reconstructing a bold quality of tone from its affiliated harmonics was early carried to a high degree of development. Indeed this curious occurrence justifies the remark in the introductory chapter that instrumental appliances intended originally for effect, and discovered fortuitously, have in the end proved important corroborations of scientific research. In the organ in the church of St. Sophia at Dresden built by Silbermann about 1750, there are besides a cornet of five ranks, a "cymbel" of similar character of three ranks, and a mixture of five ranks. In the pedal organ of the same fine instrument is a cornet of no less than eight ranks, the builder evidently knowing by instinctive experience what has now been proved by theory, that the upper partials of the deepest notes would carry with them a larger train of attendant vibrations still within the range of musical sound, than those more acute.

In the church of St. Bernhardin, at Breslau, is an organ built by Casparini in 1705, containing two cymbels, each of two ranks, and two mixtures of four and five ranks respectively. In the church of St. Mary Magdalen, in the same town, is another by a different builder, dated 1725, containing no less than eight such compound stops. The same principle is however more distinctly to be traced in what are termed mutation stops, which though composed of single pipes do not give a sound corresponding to the note pressed, but which sound either the fifth or the third above it, the first being called quint stops, the second tierces. Other forms such as the tenth, twelfth, and larigot, the last sounding a nineteenth above the diapasons, are constantly met with.

The *Quintaton* or *Quintadena*, in addition to the fundamental tone, sounds also, softly but distinctly, the twelfth above, as its name implies. It is a single stopped metal pipe of small bore, in which the first possible upper partial is unusually predominant.

The *Quint* or *Diapente*, in spite of the similarity of its name to that last given, is of a different character. That belonging to a diapason of 32 feet long is itself $10\frac{2}{3}$ in length; for an open diapason of 16 feet it is $5\frac{1}{3}$; in each case it will be seen that it gives the twelfth of the foundation tone, the second upper partial, or the fifth above the octave, whence its name. It is never drawn without its corresponding lower stop, and evidently acts by enriching the quality of this latter by auxiliary harmonics. The $10\frac{2}{3}$ stop can however be used in a somewhat different manner, by being drawn with a 16 foot open diapason. The result is very remarkable; for a grave harmonic is thus suggested to the ear, which produces by difference of vibrations an artificial 32 feet tone without any pipe of this calibre being present.

Correction of Bernoulli's Law.—It has long been known that if an open pipe be stopped at one end, its note is not exactly an octave below that given by it when open, but somewhat less; the difference being about a major seventh instead of an octave.

In a cylindrical tube open at both ends of length $= l$ and diam. $= 2r$.

Effective length $= l + 2a$.

a being the correction for one open end. If a flat stopper be applied at one end it is equal to a pipe of length $2(l + a)$. The ratio of the notes therefore is

$$(l + 2a) : 2(l + a);$$

or,

$$\frac{1}{2} \times \frac{l + 2a}{l + a}.$$

Mr. Bosanquet in investigating this subject experimentally, took an organ pipe 9·5 in. long, and ·95 in. diameter. It being difficult to obtain octaves from the same pipe open and stopped, the octave and twelfth, or second harmonic were compared. This was found to be 2 commas sharper than it should be, or an interval of about 40 : 41.

The correction for the mouth was determined by sawing across a similar pipe.

The value of a was found to be ·635 R for the open end of the pipe, and ·59 r nearly for the mouth.

He remarks that in Bernoulli's theory of organ-pipes, the hypothesis is made that the change from the constraint of the pipe to a condition in which no remains of constraint are to be perceived takes place suddenly at the point where the wave-system leaves the pipe. It is however evident that the divergence which takes place may be conceived of as sending back to the pipe a series of reflected impulses instead of the single reflected impulse which returned from the open end of the pipe according to Bernoulli's theory, and that these elementary impulses, coming from different distances, may be altogether equivalent to a single reflected impulse from a point at a little distance from the end of the pipe.

Instruments of Music proper.—In speaking of the innumerable methods of eliciting sound, it was stated that some, for musical purposes, had superseded the others. The harp, pianoforte, organ, and harmonium need not be more minutely described in a work like the present. It seems, however, advisable to give a few details as to the various instruments used in the orchestra.

The combination of sound-producers, forming a full band, is a comparatively modern invention; more so indeed than the organ. The individual instruments were many of them known and used separately; but their union into an organized body is far more recent. No such conjunction existed in classical times. It has been shown that ancient music was purely melodic, and that even vocal harmony arose from the polyphonic construction of the middle ages. Even down to Handel's days, stringed instruments, combined with oboes, were exclusively employed, occasionally reinforced by bassoons and trumpets. Haydn began to use a larger band. Mozart may be said to have originated the present arrangement. Beethoven expanded and strengthened it, and has left on record his views as to its proper size and constitution. Handel died on April 21st, 1759. The first Handel commemoration was in 1784, in Westminster Abbey. The band consisted of :—

49 first violins.	13 first oboes.	12 trumpets.
46 second violins.	13 second oboes.	6 "sacbuts."
26 violas.	6 flutes.	12 horns.
21 violoncellos.	26 bassoons.	4 drums.
15 double basses.	1 contra fagotto.	

The sacbuts were the instruments now termed trombones, being bass trumpets; then almost obsolete but now abundantly used.

At the Handel Festival in 1871 the band consisted of :—

93 first violins.	8 flutes.	6 cornets.
72 second violins.	8 oboes.	6 trumpets.
56 violas.	8 clarinets.	12 horns.
58 violoncellos.	8 bassoons.	9 trombones.
57 double-basses.	1 contra-fagotto.	3 ophicleides.
		2 serpents.
336 stringed instruments	33 wood instruments.	38 brass instruments.

Drums 8; making 415 in all.

The wind instruments are seen to have diminished materially in relative number to the strings : forming 21·1 per cent, or about ⅕ of the whole force.

Passing from exceptionally large aggregations of players such as these, it may be stated generally that the full band as now constituted consists (1) of stringed instruments in four varieties : violins, violas, violoncellos, and double-basses. Besides these there are (2) wind instruments either of wood or of brass. In the former list are usually 2 flutes, 2 oboes, 2 clarinets, and 2 bassoons. Among the brass instruments are 2 trumpets, 2 or 4 horns, and 3 trombones.

Percussion instruments are added to these, namely, kettle-drums, side-drums, and cymbals.

Other wind instruments only occasionally employed are the piccolo, or octave flute, the cor anglais, or tenor oboe, the bassethorn, or tenor clarinet, the contra-fagotto, or double bassoon, among the wood-wind instruments ; and the cornet, the ophicleide, the euphonium, or bass saxhorn, the saxiphone, the sonorophon, and others of less importance.

The violin family **consist** essentially of a hollow resonant **case of** peculiar form, strung with four catgut strings, and **excited** by means of a rosined **bow.** They are first explicitly mentioned in Zacconi's *Pratica di Musica*, published in 1596, though they **were** known previously in this country, and used in the royal band.

They **are** the **descendants** of **the** viol family, which, however, had six strings instead **of** four, and *frets*, a contrivance now disused except on the guitar, by which the length of string for each particular note was permanently fixed, and not as now determined by the finger and the ear of the performer. Viols were made in four sizes termed :—

Descantus.	Altus.	Tenor.	Bassus.
Dessus.	Haut Contre.	Taille.	Basse.

These names **are** evidently formed after the corresponding quality of **the** human voice.

The only viol still in use is the double-bass, which differs both **in** shape, number of strings, and mode of tuning from **the** violins proper. A concert of viols was however revived at the Antient Concerts **on** the 16th of April, 1845, in which the whole **set** or " chest " of them was employed, at the request of the late Prince Consort.

The Violin has four strings tuned in fifths, and named downwards as follows:—

First string tuned in E.
Second ,, ,, A.
Third ,, ,, D.
Fourth ,, ,, G.

The intervals between these notes are obtained by shortening the string by pressing it with the tips of the fingers of the left hand on the prolonged "finger-board." The lowest note is obviously the G of the 4-foot or small octave, the seventh or 2^7, in Mr. Ellis's notation. The upper limit is less distinctly defined, since, by means of harmonics, very acute notes can be produced.

The Viola, or tenor violin, more properly termed the alto abroad, is a violin somewhat larger in size, with the highest string removed, and replaced by another below, at the same interval of a fifth, therefore giving C of the 4-foot octave. Its strings are as follows:—

First tuned in A.
Second ,, ,, D.
Third ,, ,, G.
Fourth ,, ,, C.

The quality of the viola is more hollow and mournful than that of the violin.

The Violoncello is an octave below the viola. Its four strings are tuned to the octave of those on the former. It consequently reaches down to the C of the great, or 8-foot octave, the sixth or 2^6 of Mr. Ellis's notation.

The Double Bass is said to have been invented by Michele Todini at Rome in 1676, but it probably existed in some form long before this date. Being essentially a viol it has a different shape from its nearest kindred the violoncello. It usually has three strings, and is tuned in fourths instead of in fifths. The notes are as follows:—

First string tuned in G.
Second ,, ,, D.
Third ,, ,, A.

It thus reaches only two real notes below the violoncello, that is to the A of the contra or 16-foot octave the 2^5 or fifth of Mr. Ellis's notation. Its larger body of tone makes it seem of a deeper pitch than that it actually possesses.

In Germany a fourth string, tuned to E, has long been used, a fourth lower than that of the ordinary instrument. But it is perfectly easy and most desirable to carry the pitch still lower to the C, or foundation-note of the 16-foot octave. In this part the band is decidedly weak, and far inferior to the organ or even the pianoforte. Beethoven must have felt this defect, for he writes double-bass parts, such as those in the Pastoral and C minor symphonies, down to this note. A double-bass strung down to contra C was exhibited by the writer at a recent International, and at the Loan Exhibitions. The principle has been already described.

Orchestral Wind Instruments have already been described as regards their principle. Those in actual musical use are of three kinds.

1. Flutes.
2. Reeds.
3. Instruments with cupped mouthpieces.

They all require two essential parts: I. the windchest; II. the embouchure.

The windchest in this case is invariably the human thorax. The writer made a series of experiments some years ago for the purpose of determining what the pressures within the thorax actually were. A water gauge was connected with a small curved piece of tube by means of a long flexible india-rubber pipe. The curved tube being inserted in the angle of the mouth, did not, after a little practice, interfere with the ordinary playing of the instrument. The various notes were then sounded successively, and the height at which the column stood was noted. The following table of pressures was obtained as an average of many experiments :—

Table of Pressures.

Oboe	9	inches to	17.
Clarinet	15	,, ,,	8.
Bassoon	12	,, ,,	24.
Horn	5	,, ,,	27.
Cornet	10	,, ,,	34.
Trumpet	12	,, ,,	33.
Euphonium	3	,, ,,	40.
Bombardon	3	,, ,,	36.

172 ON SOUND. [CHAP.

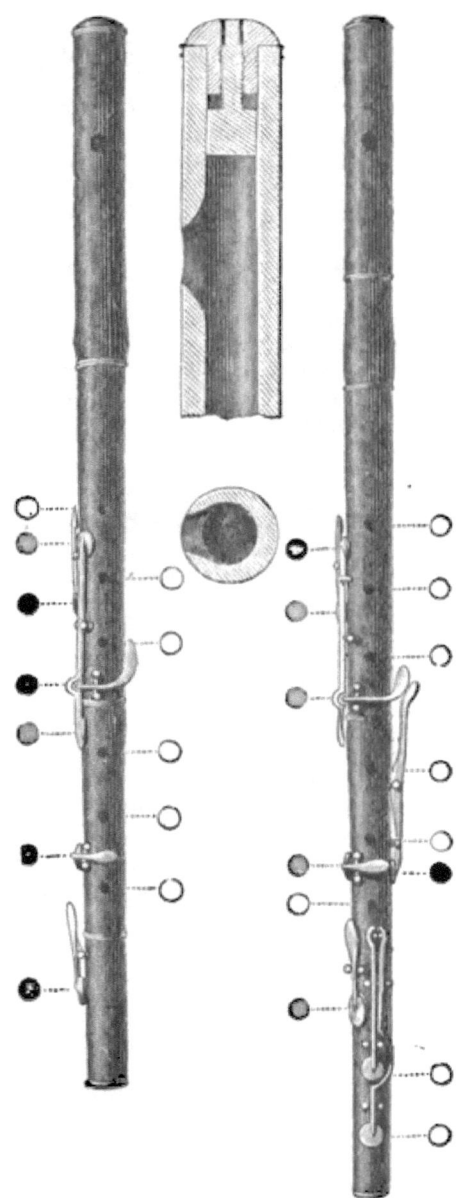

Fig. 57.—The flute. Longitudinal and transversal section of the mouthpiece.

The Flute is an instrument of great antiquity, but not in the form in which it is now played. It acts like the ordinary open organ pipe, by driving a current of air from the lips against a thin edge. This edge in the modern form is fashioned in the side of a large lateral hole near the upper extremity. In the olden form, that of the flute à bec, or flageolet, there is a fixed contrivance like the mouth of an organ pipe for producing the tone, and the wind is simply blown into the mouthpiece. There is, however, no reason why either of these systems should be adopted. The writer has particular pleasure in drawing attention to a reed flute brought from Egypt by his friend Mr. Girdlestone, of the Charterhouse, which exactly illustrates the stage of development of this instrument hitherto wanting. It is about fourteen inches long, possessing the usual six finger-holes, but the upper extremity or head is continuous. The top end is not stopped with a cork, as in the ordinary flute, but is thinned off to a feather edge, leaving a sharp circular ring at right angles to the axis of the bore. If this flute be held obliquely towards the right hand of the player, and the stream of wind from the all but closed lips directed against the opposite edge of the ring, a fair but somewhat feeble flute tone can be elicited.

Fig. 58.—Nay or Egyptian Flute.

Here the mechanism is reduced to its very simplest form. It is moreover interesting to observe that the flute still played on by the peasants about the Nile is the counterpart of that to be seen distinctly on the Egyptian Hieroglyphics of many thousand years ago.

The Oboe or Hautbois is one of the very earliest instruments known; Mr. Chappell has succeeded in reproducing exact copies of real specimens found in the Egyptian tombs, and the writer has fitted reeds to them, by means of which a fair musical scale can be elicited. Beside the originals in the tomb lies usually a small piece of grass or reed, obviously intended to furnish the means of playing. In some cases this has been actually within the bore.

The "reed," or "cane" now used for all reed instruments is formed of the outer siliceous layer of a tall grass the *Arundo donax* or *sativa*, which grows in the south of Europe. This is fashioned in the oboe and bassoon into a broad spatula-like form with two thinned plates of the cane in close approximation to one another.

It is therefore **termed a double reed**; in opposition to that of the clarinet and some other instruments where the vibrating plate of cane is single. It has been materally reduced in size of late years with a corresponding improvement in the tone of the instrument. Even as late as the visit of the composer Rossini to this country a reed resembling that of the bassoon was in use for the oboe.

The bore of the oboe is conical, enlarging at the lower extremity into an expanded bell. Its scale is founded on the interval of the octave, beginning at the $B\flat$ or $B\natural$ of the four foot or small octave, and extending to F in alto in the twice accented, or one foot, or 9th octave.

The Clarinet is an instrument of **four-foot tone**, with a single reed and smooth quality, commonly said to have been invented in 1690 at Nuremberg. It is probable, however, that in one form or another it existed long before. Its name is evidently a diminutive of *Clarino*, the Italian name of the trumpet, to which its tone has some similarity.

The clarinet consists of a peculiar mouthpiece furnished with a **single beating** reed, a cylindrical tube terminating in a bell, with eighteen openings in the side, half of which are closed by the fingers, and half by keys. The lower scale comprises nineteen semitones, from F in the base stave to $B\flat$ in the octave above. The lowest note is emitted through the bell, the G of the two-foot octave through a hole at the back of the tube, peculiar to this instrument. This register is termed *Chalumeau*. By opening a lever above the back hole named above, the pitch is raised a twelfth, so that the E of the small or four-foot octave becomes the $B\natural$ of that above. By the successive removal of fingers, fifteen more semitones are obtained, reaching to high $C\sharp$, and above this note is another octave obtained by cross-fingering.

The mouthpiece is a conical stopper flattened at one side to form the table for the reed, and thinned to a chisel edge on the other for the convenience of the lips. From the bore a lateral orifice is cut into the table which is closed in playing by the thin end of the reed. The table on which the reed lies, instead of being flat, is curved backwards towards the point, so as to leave a gap or slit about the thickness of a sixpence between the end of the mouthpiece and the point of the reed.

Helmholtz has analyzed the tone and musical character

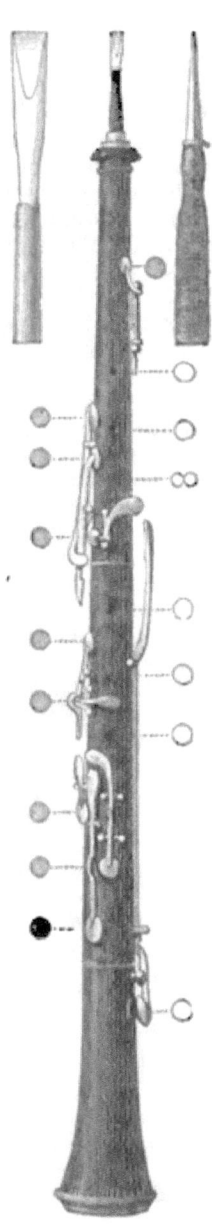

Fig. 59.—Hautbois.
Front and side view of reed.

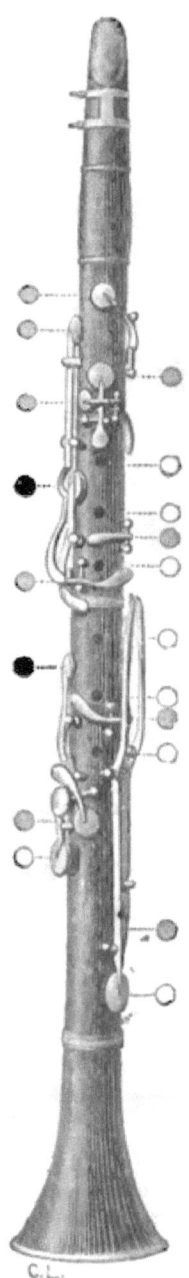

Fig. 60.—Clarinet.

of the clarinet, as has been stated above. It stands apart from all other instruments, both in its quality, in its scale, founded on the twelfth, and according to the writer's experiments, in the wind-pressure required for its various registers. The clarinet is made in many keys, to meet difficulties of execution. This fact enables a very close approach to true intonation to be obtained on it, as described in the chapter on Temperament.

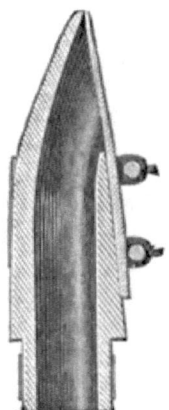

Fig. 61.—Section of mouthpiece.

The Bassoon is a double-reed instrument of eight-foot tone, as implied in its name; it being the natural bass of the oboe. In one form or another it is probably of great antiquity; though it is said to have been invented in 1539 by Afranio, a canon of Ferrara. A class of instruments named *bombards*, *pommers*, or *brummers* seem to have been the immediate predecessors of the bassoon. It is a contrivance which has evidently originated in a fortuitous manner, developed by successive improvements of an empirical character. Various attempts have been made to give greater accuracy and completeness to its singularly capricious scale, but with only partial success. Its compass is from $B\flat$ in the contra or sixteen foot, to $A\flat$ in the once accented or two-foot octave, but additional mechanism has greatly raised the upper limit, so that the C or even the F above that note can be obtained.

Like the oboe, it gives the consecutive harmonics of an open pipe.

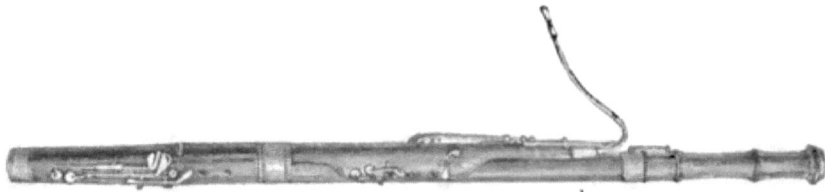

Fig. 62.—Bassoon.

Instruments with cupped mouthpieces may be cited as the simplest musical instances of consonant tubes. They all consist essentially of an open conical tube, often of great length, in the French horn about 17 feet. The fundamental note of such a tube is consequently very deep. At the

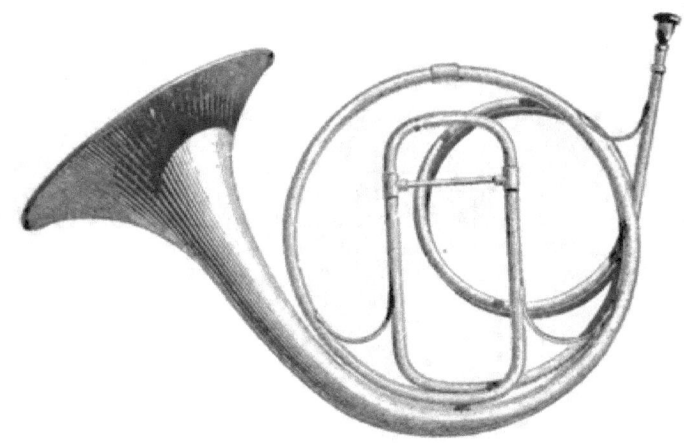

Fig. 63 —French Horn.

smaller extremity is the cup, forming an expansion of the bore, carrying a rounded edge against which the tense lips of the player are steadily pressed. The reed thus constituted is of the membranous kind, not dissimilar to the vocal chords of the human larynx. The method of its vibration is totally different from the reed of the oboe or clarinet: for whereas in these the lower harmonic notes are damped by the appended tube, and one of the higher and sweeter partials is reinforced; in the cupped instruments every successive harmonic from the very lowest is practicable, and all but the extreme bass sounds are actually used successively in producing the scale. The sequence of sounds is the harmonic series already given, modified slightly according to the particular instrument; it depends for its production entirely on the varied tension of the lips, and is commonly termed the scale of Open Notes. It is to bridge over the long gaps and intervals between these open notes that all systems of valves, slides, and keys are intended. The natural or open notes are as follows, in the French Horn, which furnishes the most perfect example of the class :—

178 ON SOUND. [CHAP. VIII.

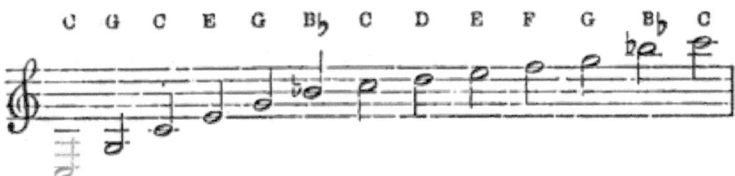

It will be seen that in the lower part of the series the intervals between the sounds are large, but that the upper harmonics approach closer and closer together, so that from the middle B♭ a nearly perfect octave scale of continuous notes can be obtained. It has long been the custom to interpolate the missing semitones on the French horn by thrusting the hand into the bell, and so lowering the pitch by a variable quantity. The instrument is hence named the "hand" horn, and the notes so modified hand-notes. Of late years, however, valves have been applied, as will be described in a subsequent paragraph.

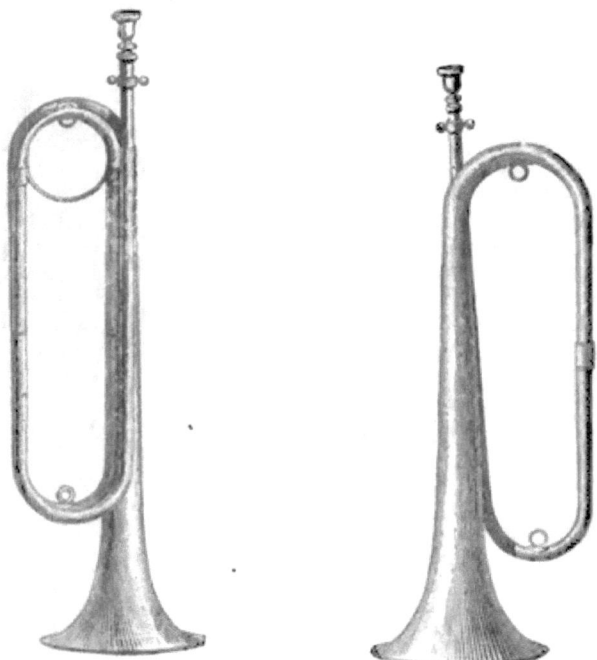

Fig. 61.—Trumpet and clarion.

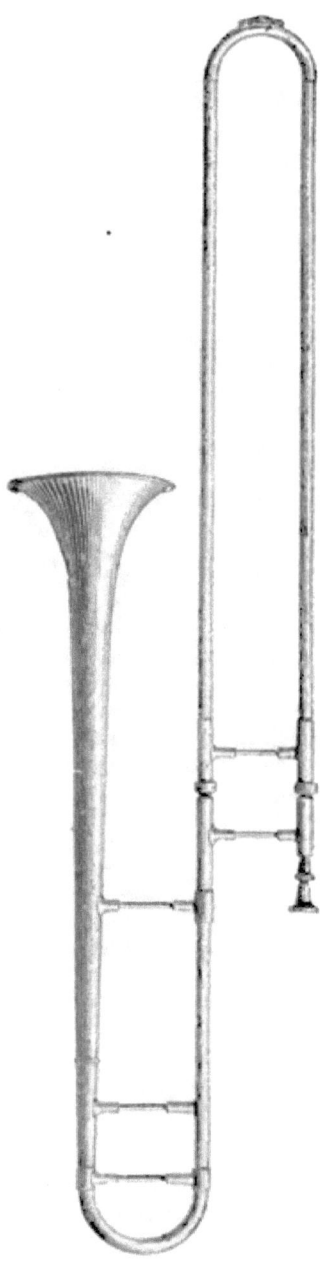

Fig 65.—Trombone.

The Trumpet.—Speaking in a higher octave possesses the first eleven open notes of the French horn. In this instrument, and in the trombone, its natural bass, a totally different, and far more perfect system has been adopted for completing the scale. An U shaped portion of the tube is made to slide with gentle friction upon the body of the instrument, so that the length of the bore can be increased and diminished by any given quantity within certain limits, at the will of the player. The note emitted can thus be lowered insensibly, and without abrupt changes through a variable interval. The absence of fixed notes enables the intonation to be guided at the will of the player, by accurate ear, exactly as is the case on the violin family. We therefore have in the trumpet and trombone quartett a perfect combination as regards temperament, and one equally well diversified by contrast of tone. It is to be regretted that owing to the difficulty of finding competent players, this most perfect department of the wind-band is falling into neglect and disuse.

The Ear is divided into three parts, the external, middle, and internal. The latter contains the nervous apparatus concerned in hearing; the two former act as conveyances of aerial vibrations: they are not absolutely necessary to the act of hearing, for persons suffering from deafness due to disease of the middle ear can often hear a musical note by applying the sounding body, such as a tuning-fork, to the adjacent bones of the head, through which the vibration is transmitted.

The external ear consists of the auricle or *pinna*, and the auditory canal, or *meatus*.

Considerable doubt exists as to the part played by the auricle in the act of hearing. That it has some influence is clearly shown by the fact that in the lower animals it is movable, and is instinctively directed towards the source of sound. Hence is derived the common phrase "Pricking up the ears." In the human subject the muscles which actuate it are all but atrophied, and the power of motion is slight and unusual. The ears however of such animals as the horse, cat, and ass are of a trumpet shape, and eminently fitted to discriminate the direction of sonorous undulations. Even in man it retains some such power, which is often assisted unconsciously by supplementing it with the hand, and bending it forward into a more efficient position. It retains the funnel-shape of the lower animals to a considerable degree.

The meatus is about an inch and a quarter long, directed slightly forwards as well as inwards, closed somewhat obliquely at its inner extremity by the tympanum or drum of the ear.

The middle ear is an air-containing cavity, having communication with the atmosphere by means of a tube named the Eustachian, which passes from it to the upper part of the throat.

The tympanic cavity contains three small bones termed respectively the malleus, incus, and stapes, or the hammer, anvil and stirrup, from fancied resemblances of their shape

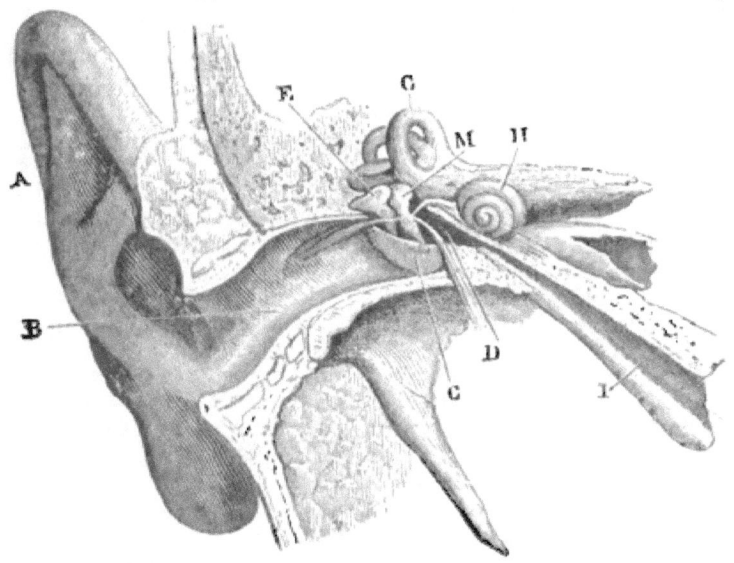

Fig. 66.—The human ear; section of the interior tympanum; chain of small bones. Internal ear; labyrinth.

A Auricle; B External Meatus; C Tympanum; D Middle Ear; E Incus; M Malleus; G Semicircular Canals; H Cochlea; I Eustachian Tube.

to these objects. The three bones are connected by small articulations, and are attached at the two extremities of the chain; the handle of the malleus being fixed to the inner side of the tympanic membrane somewhat below its middle, and the base of the stapes to the fenestra ovalis, an aperture of oval contour in the bony wall of the internal ear or labyrinth. Three muscles act on these small bones, two attached to the malleus, one to the stapes; the two former being the tensor and laxator, of the tympanum, the third, the stapedius, acting on the neck of the latter bone. The tympanum communicates with cells also filled with air, hollowed out of the mastoid process of the temporal bones and hence called mastoid cells.

The internal ear is entirely embedded in the hard or petrous portion of the temporal bone. Its complicated and anfractuous contour is lined with a membranous sac. It is not inaptly termed the labyrinth, and is divided into the vestibule, the cochlea, and the three semi-circular canals. The outer wall of the vestibule is perforated by another opening besides the fenestra ovalis, namely the fenestr᷉ rotunda. The inner wall has a number of small holes transmitting branches of the auditory nerve, the *Portio mollis* of the seventh cranial nerve. The two fenestræ thus communicate with the middle ear, the smaller orifices with the inside of the cranium.

The semicircular canals communicate by five openings with the upper and posterior part of the vestibule; two extremities, those of the superior, and of the posterior canals entering by a common termination. Each canal is expanded at one end into a globular enlargement termed an ampulla.

Besides the superior and posterior, the third canal is named from its position, the external. They stand in three rectangular planes to one another; thus representing the three dimensions of space.

In front of the vestibule, in form of a prolongation, is the cochlea, similar in shape to a snail-shell, a flat cone with apex outwards, consisting of a spiral taper tube of two and a half turns around the axis. The central column of bone sends a partition, named the lamina spiralis, outwards from its centre, similar to the thread of a taper screw, but defective at the apex. This lamina of bone is met and completed by two membranous prolongations termed the basilar membrane and the membrane of Reissner.

In this way three helical passages are cut off, termed severally the *scala tympani*, the *scala vestibuli* and the *ductus cochlearis*, the last being the smallest and the least distinct. The two former are connected at the apex, and at the bottom end, one, the scala vestibuli, with the vestibule, and the other, the scala tympani, with the interposition of a membrane, opening by the fenestra rotunda into the tympanic cavity.[1]

The membranous labyrinth lies freely in the bony cavity, except as regards the ductus cochlearis, which is in close approximation to the bony wall of the labyrinth. Fluid is contained in its sac, and also between it and the wall. The membranous lining of the vestibule is divided into two non-communicating parts, the larger termed the *utriculus*, the

---

[1] These modes of describing the "canalis membranacea" and organ of Corti are judiciously adopted by Mr. Champneys in Stainer and Barrett's *Dictionary of Musical Terms*.

anterior and smaller the *sacculus*, which latter opens into the ductus cochlearis.

Within the membranous canal or cochlear duct, separated from it by a layer of membrane, lies what is named after its discoverer the *organ of Corti*.

This remarkable apparatus is placed between a lower or basilar membrane, stretched over the scala tympani from the bony lamina spiralis to the outer wall, and the membrane named above, separating it from the cochlear duct. It consists of a double series of diminishing rods, following the spiral of the cochlea, about 3,000 in number. Their bases are separate, but their upper ends meet at an angle, "forming a sort of spiral gable roof." It has latterly been supposed that these are the instruments by which we distinguish the pitch of sounds. Each one is considered to vibrate in sympathy with one note, and to transmit its vibrations to a special twig of the nerve of hearing. The semi-circular canals are supposed to enable us to determine the direction of sounds, a theory which their relation to the three dimensions of space seems to justify. But the point cannot be considered as finally settled.

Education of the Ear.—It is somewhat singular that in the majority of acoustical researches the eyes have been made to do more service than the ears; and when the eyes have failed, that they have been replaced by calculation or theory. One result of this curious fact is that since the discovery of ingenious methods for rendering the vibratory motions of solids visible to the eye, many facts have been substantiated as to the vibrations of strings, plates, and rods, but far fewer as to those of liquids and gases. Indeed, the movement of the air in an organ-pipe is still matter of dispute, although it may be studied optically by means of a very simple contrivance.

It is not, however, sufficiently appreciated that the education of the ear may render that organ competent to undertake higher functions. In this respect the ear resembles all the other senses.

In many cases it is simply aural memory which needs cultivation. Even an ordinary countryman can recognize the sound of his village bell after twenty years' absence; just as we all recognize a friend whom we have not seen for long, and whom time's ravages have made a stranger to the eye, by the sound of a single spoken word. To this category belongs the tonal memory which distinguishes absolute pitch. There can be no doubt that the faculty exists in some persons to a very high degree. It is an entirely different question whether it is what is commonly termed a "gift" or an acquirement.

In a great degree, doubtless, it is mixed with what has been lately recognized as the sixth sense, namely, the "muscular sense;" by which we appreciate the amount of nervous influence or muscular contraction necessary to produce a certain result. Just as, by means of the "muscular sense," we can approximate to an estimate of weight or speed, so by the instinctive contraction of the laryngeal muscles we mentally produce a unison to any heard sound, and, as it were, weigh it in the balance against our own vocal powers. That this is the real process in many cases is evidenced by the fact that the fortunate owners of clear, accurate voices, are specially endowed with the power in question. But it is not confined to these; and, in some instances, appears to be a pure act of memory. But it will usually be found to exist in this form in old practised musicians, whose piano, organ, or other musical instrument, has become, by long familiarity, almost a part of themselves. Another frequent indication of its acquired nature, as against its innate possession, is the fact that different forms of acuteness can by practice be developed in ears of otherwise equal sensibility. Where the glee-singer, or chorus-master, or organist, acquires the power of giving the key-note, the piano- or organ-tuner cultivates a sensibility to beats, or differences of tone, which the former hardly hears at all. Upon this point the observations of Prof. Preyer, quoted by Mr. Ellis, are of very great value.

The Sensibility of the Ear for the appreciation of weak sounds is very great. Every fly which buzzes around gives to a musical ear the power of instantly testing the number of vibrations per second which its wings go through. Even between the limits of high to deep vibrations the ear has vast latitude, from 32 in a second to ten thousand, or in exceptional cases far beyond. The eye possesses at most an octave of sensations, but the average ear easily runs through seven or eight.

Even in the matter of accuracy, a comparison between the eye and ear ends in favour of the latter. An architect or draughtsman who between two lines, neither parallel nor in one plane, made an error of estimation by eye not exceeding $\frac{1}{30}$ would gain credit for unusual precision. But in the ear $\frac{1}{30}$ amounts to a quarter of a tone, and by ear $\frac{1}{45}$ of a tone is easily determined. Still more remarkable is the ear in the rapidity of its appreciations. They are all but instantaneous. A sound lasting only a quarter of a second is as well or better gauged than if it lasted a minute, and in this short period if several sounds be simultaneously produced, the ear can instinctively separate them without any risk of confusion. It is a common mistake to suppose that what is termed a "good ear" is a rare gift. No doubt a cultivated organ, competent to distinguish musical sounds with accuracy, is far from usual. But from experiments made by Marloye, Duhamel, and the writer, it may be safely affirmed that the defect is not in the organ of hearing itself. Marloye experimented on a large number of persons, without finding a single defective case. There were many perfectly devoid of any practice in estimating sounds, and great differences in the amount of delicacy they could acquire; but all without exception were, he states, susceptible of instruction. He shows that it is of the utmost importance that this instruction should be begun early in life, for in childhood the sensitiveness of the organ of hearing increases rapidly with use.

Preyer's Researches.—Dr. W. Preyer, Professor of Physiology in the University of Jena, has made valuable experiments on this subject. He uses a differential apparatus consisting of 25 harmonium reeds, the first 10 proceeding from 500 to 501 vibrations, by tenths of a vibration, and the remainder tuned to 504, 508, 512, 1000, 1000·2, 1000·4, 1000·6, 1000·8, 1001, 1008, 1016, 1024, 2048, 4096 vibrations per second.

It appeared from trials with this instrument, that taking two notes in rapid succession, there were two questions: (1) Are the tones different? (2) Which is the sharper of the two? Many ears can detect the difference which cannot decide the second question.

The results showed that in the 8-foot octave, a difference of ·418 vibrations per second, in the 2-foot octave ·364 vibrations, in the $\frac{1}{2}$-foot octave ·500 vibrations, could be detected. Speaking generally, throughout the scale a difference of $\frac{1}{2}$ of a vibration is not heard; whereas, in the contra or 16-foot octave $\frac{2}{3}$ of a vibration; in the 2-foot octave $\frac{1}{3}$; in the higher registers $\frac{1}{2}$ can be detected.

In extremely low notes the accuracy of appreciation is considerably less, and requires special practice. The same is true, though in somewhat different degree, of the upper extremity of the scale.[1]

The Larynx.—The organ of voice is essentially a reed instrument of peculiar construction, consisting of two semicircular elastic membranes, approximated at their straight edges, but leaving a narrow crack between them, through which air can

---

[1] For farther details see *Proceedings of Musical Association*, 1876-7. Paper by Alexander J. Ellis, Esq. "On the Sensitiveness of the Ear to Pitch and Change of Pitch in Music."

be forced from the lungs with a view to set them into vibration. The solid portions of the frame in which these membranes, erroneously termed vocal chords, are set, consist of cartilage.

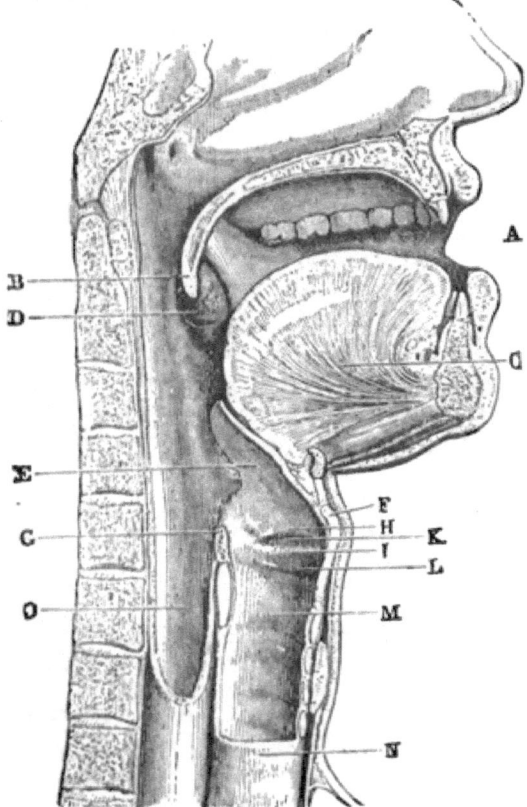

Fig. 67.— The human voice; interior view of the larynx. Glottis; vocal chords.

A Oral Cavity; B Soft Palate; C Tongue; D Tonsil; E Epiglottis; F Thyroid Cartilage; G Arytænoid Cartilage; H I Vocal Chords (Superior and Inferior); K Ventricle; L Rima Glottidis; M Cricoid Cartilage; N Trachea; O Æsophagus.

They are nine in number, three symmetrical, and occupying the middle line of the body, namely, the epiglottis, the thyroid, and the cricoid cartilages; six in pairs, namely, the two arytenoid cartilages, and two smaller accessory cartilages, named after Santorini and Wrisberg.

The thyroid cartilage forms the upper part of the hard mass to be felt in the neck, and commonly called "Adam's apple." It rests upon the cricoid cartilage, with which it is articulated at either extremity. It is in shape somewhat like a partly-opened book with the back directed forwards. At the posterior edges are four horns, the **lower pair** connected, as named above, with the cricoid, the upper pair suspended to the hyoid bone at the root of the tongue.

The cricoid cartilage is not unlike a signet ring, with a broad facet **directed** backwards, upon which are two pairs of smooth surfaces, the lower pair for the attachment of the horns of the thyroid, the upper for the loose articulation of the two arytenoid cartilages.

The arytenoid cartilages are of an irregular triangular or pyramidal shape, with their apices upwards, their bases resting on the broad part of the cricoid ring. Two other small cartilages, those of Santorini, are set on the apices of these.

The epiglottis is an oval leaf-shaped flexible plate of cartilage, situated at the back of the tongue. The tail or stalk is attached just within the notch of the thyroid body, and the front is connected by an elastic band with the back of the hyoid bone.

The vocal chords themselves are attached in front just below the tail of the epiglottis, **to the** angular opening between **the sides of the** thyroid cartilage. This attachment is there**fore fixed. Their** posterior attachment is to the apices of the arytenoid cartilages, and is therefore free **to move** in two directions. When the arytenoid cartilages **are** drawn backward by muscular action, a tension **is** put upon the vocal chords; **when** they are drawn outwards or inwards, the back extremities of **the** vocal chords are either widely separated **or** closely approximated. From this approximation and tension the vibration and consequently the musical note of the voice is produced.

It has become possible **to give a** much more satisfactory description of the position and action of the vocal chords since the invention of the laryngoscope, a simple but effectual reflecting apparatus, by means of which the interior of the mouth and larynx can be examined. It must be remembered that the epiglottis is the point at which the respiratory and alimentary passages cross or decussate.[1] The proper orifice of the former is the nostril, whence the direction, at first horizontally backwards, turns downwards behind the veil of

---

[1] The two organs and their appendages are physiologically distinguished by **the** difference of the epithelium, that of the respiratory tract being ciliated, **that** of the alimentary, tesselated.

the palate. The opening of the food passage is the mouth, whence it passes over the epiglottis to the posterior fauces, the œsophagus lying behind the trachea. During swallowing the laryngeal opening is tightly closed, and it is astonishing how accurately the mechanism for preventing the entrance of food into the trachea works.

During ordinary breathing the epiglottis lies back and the tips of the arytenoid cartilages can be seen with the laryngoscope. The opening of the glottis is diamond-shaped. In pronouncing the vowels it rises and shows the larynx, the vocal chords being brought together. For the production of low chest notes, the arytenoid cartilages bend under the overhanging epiglottis; but during the utterance of high chest notes, it is easy to see them, with the vocal chords closely approximated. "In shrill cries the cushion of the epiglottis appears to be pressed down on the front part of the vocal chords so as to shorten their vibrating portion, just as a string is shortened by the fingers on the finger-board of a stringed instrument."[1]

The name of chords obviously suggests an erroneous hypothesis as to their function. No string so short could produce so low a note as those of the male voice, and the laws of string-tension are not observed. Neither can they be supposed to act on the principle of a diapason pipe. They fulfil far more closely the conditions of a free reed of a membranous character with a double vibrator. They possess a tube below them in the trachea and bronchi, with a complex resonance tube above them in the cavity of the mouth, pharynx, nose, and frontal sinuses. This is farther described in speaking of the vowel sounds. Helmholtz considers the *head voice* to be produced by a thinning of the edge of the chords by drawing aside the mucous coat below them. The production of the *falsetto* is still open to considerable doubt, some holding that the chords only vibrate in part of their length; others that they vibrate in segments, giving harmonics; others again ascribe them to the same cause as that of the head voice, and they have been attributed to the folds of mucous membrane above them, called *false* vocal chords. Perhaps the most probable theory is that of harmonics, although the exact mechanism is unknown.

[1] Czermak, as quoted in Stainer's *Dictionary of Musical Terms*, Article "Larynx."

INDEX.

A.

Air, variation of density in, 53
    elasticity of, 54
    velocity of sound in, 48
    motion of, in pipes, 39
Amplitude of vibrations, 62
Analysis of musical tones, 113
Appunn's reed tonometer, 102
Atmosphere, refraction by, 60
Atmospheric pressure, effect of, 128
Auditory nerve, 182

B.

Barometric pressure, effect of, 131
Bars, vibration of, 15
Bassoon, 176
Beats, 72
    of upper partial tones, 124
Bells, 27
Bernoulli, his laws, 40
    correction of, 167
Blake's photographic method, 99
Bosscha's experiment, 54

C.

Cagniard Latour's siren, 81
Causes of rise in pitch, 108
Chladni—his tonometer, 104
    his sand figures, 26
Chemical harmonicon, 43
Chords, 137
Clarinet, 174
Clarke and MacLeod's tonometer, 94
Columns of air, vibration of, 36
Combinational tones, 123
Combination of vibrations, 88
    plate of, 92

Compound nature of musical tone, 117
Consonance, 65
Corti, organ of, 183
Cupped mouth-pieces, 176

D.

Development of scale, 120
Diatonic scale, 139
Difference tones, 124
    of intervals, 126
Dissonance, 4
Distributors of sound, 2
Double bass, 170

E.

Ear, physiology of, 180
    education of, 183
Echoes, 58
Edison's phonograph, 89
Effects of heat, 127
Electricity a source of sound, 5
Equal temperament, 144
Error of siren, 83
Explosion a source of sound, 4

F.

Fiddle, 170
Flame—singing, 42
    sensitive, 44
    manometric, 94
Flute, 173
    Egyptian, 37
Fog, influence of, 161
Fourier's theorem, 112
Free reeds, 85
French horn, 176
Friction as a source of sound, 4

INDEX.

G.

Gases, velocity of sound in, 49
    reflection from, 59
Graphic methods, 85

H.

Harmonicon, 20
Harmonics, 118
Harmony, 162
Hautbois, 173
Hearing, mechanism of, 180
Heat—effect on sound, 127
    evolved in vibration, 53
    effect on strings, 129
    ,, ,, tuning-forks, 130
    ,, ,, reeds, 131
    ,, ,, organ pipes, 131
Helmholtz's vowel theory, 116
Horn, 177
Hydrogen, sound velocity in, 49

I.

Intensity, 62
Interferences of sound, 70
Intervals, 135
Instruments of music, 167
Irregular vibration, 4

K.

Keyboards, 144
    Bosanquet's, 152
    Colin Brown's, 154
    Gueroult's, 148
    Helmholtz's, 148
    Poole's, 151
    Thompson, Perronet, 147
Kœnig's manometric flames, 94
Kundt, his experiments, 55

L.

Laplace, **his correction of Newton's** formula, 53
Larynx, 185
Lateral vibration of rods, 16
Laws **of** Bernoulli, 40
Limits of audible sound, **75**
Liquids, velocity in, 50
    temperature modification, 129
Lissajous, his figures, 90
Longitudinal vibrations, 15
    of strings, 13
    of rods, 16

M.

Major scale, 139
Manometric flames, 94
Marimba, 21
Mayer's electrical tonometer, **102**
Mean **tone** temperament, 144
Mechanical methods of tonometry, 77
Melde, his experiments, 12
Membranes, vibration of, 33
Metals, velocity **of** sound in, **14**
Microphone, 48
Minor scale, 142
Mixtures, 164
Moisture, influence of, **129**
Monochord, 84
Motion of air in pipes, **39**
Mouth, resonance of, 116
Music, special application **to, 161**
Musical box, 19
    glasses, 31
Musical sound and noise, 1

N.

Nail-fiddle, **17**
Nature of musical sound, 110
Newton's velocity of sound, 53
Nodes, **9**
Notation, 162

O.

Oboe, 173
Octave, the, 135
Open pipes, 39
Optical tonometry, **90**
    curves, 91, 93
Orchestra, the, 158
Orchestra, true intonation in, 153
Orchestral wind instruments, 171
Organ pipes, varieties of, 164
    quality **of, 122**
    effects **of heat on, 131**
Over tones, 111

P.

Partial **tones, 111**
Phonautograph, 87
Photography of vibrations, 99
Pipes, see Organ pipes, 165
Pitch, 75
    alteration **of,** by motion, **109**
Plates, **vibration** of, 25
Pressure, effects of, 128
Preyer's researches, 185
Propagation of sound, 46
    by earth, 46
Pyrophone, 44
Pythagorean tuning, **136**

INDEX.

Q.

Quality, 110
Quintaton, 166

R.

Rayleigh's **experiment, 103**
Reeds, 34
Reflection of sound, 56
Refraction of sound, 56
Regular vibration, 5
Relative harmoniousness of chords, 137
Resonance, theory of, **68**
Resonators, 66, 112
Resultant tones, 123
Rise in pitch, causes of, **108**
Rod, vibrations of, 15
Rods free at both ends, 19

S.

Savart's wheel, 78
Scale, 139
    Pythagorean, 142
Scheibler's tonometer, 100
Sensitive flames, 44
Shadows, sound, 59
Shock as a source of sound, 4
Singing flames, 42
Siren, 79
    double, 82
    Seebeck's, 80
Solids, velocity in, 51
Sondhaus's experiment, 42
Sonometer, 11
    for longitudinal vibration, 14
Sound, definition of, 1
Sound boards, 69
Sources of sound, 2
Standards of pitch, 105
Stopped pipes, 40
Strings, 6
    effects of heat on, 129
Summation tones, 124
Synthesis of tones, 114

T.

Tartini's tones, **3**
Telephone, Wheatstone's, **47**
    Graham Bell's, 48
Temperament, 143
    table of, 144
Thompson, Perronet,
    his organ, 147
Timbre, 110
Tonometers, 100
Toothed wheels of Savart, **79**

Torsional vibrations, 14
Transverse vibrations, 6
Trevelyan's rocker, 41
Trombone, 179
Trumpet, 180
Tuning-forks, 21
    effects of heat **on, 130**

U.

Upper partial tones, 113
    reality of, 113

V.

Variation of pitch, **105**
Velocity, of sound, **48**
    in gases, 129
    influenced by temperature, **127**
    "   "   density, 49
    "   "   elasticity, 49
    Newton's calculation of, 48
    "   "   " gases, 49, 129
    "   "   " liquids, 50
    "   "   " in solids, 51
Ventral segments, 9
    from heat, 41
Vibration longitudinal, 5
    transverse, 6
    torsional, 14
Vibration of strings, 6
    rods, 15
    plates, 25
    bells, 27
    membranes, 33
    reeds, 34
    columns **of air, 36**
Vibration **microscope, 122**
Vibroscope, 86
Viola, 170
Violin, 170
Violoncello, **170**
Voice, 185
Vowel sounds, analysis of, 116
    synthesis of, 117

W.

Water, waves in, 63
    velocity in, 50
Wave motion, 52
Waves, length of, 54
Wertheim's experiments, 54
Wires, *see* Strings, 6
Wood, velocity in, 47

Z.

Zanze, 20
Zambomba, 34

LONDON:
R. CLAY, SONS AND TAYLOR, PRINTERS
BREAD STREET HILL.

www.ingramcontent.com/pod-product-compliance
Lightning Source LLC
Chambersburg PA
CBHW021731220426
43662CB00008B/792